Synthesis Lectures on Computer Vision

Series Editors

Gerard Medioni, University of Southern California, Los Angeles, USA

Sven Dickinson, Department of Computer Science, University of Toronto, Toronto, Canada

This series publishes on topics pertaining to computer vision and pattern recognition. The scope follows the purview of premier computer science conferences, and includes the science of scene reconstruction, event detection, video tracking, object recognition, 3D pose estimation, learning, indexing, motion estimation, and image restoration. As a scientific discipline, computer vision is concerned with the theory behind artificial systems that extract information from images. The image data can take many forms, such as video sequences, views from multiple cameras, or multi-dimensional data from a medical scanner. As a technological discipline, computer vision seeks to apply its theories and models for the construction of computer vision systems, such as those in self-driving cars/navigation systems, medical image analysis, and industrial robots.

Yue Song · Thomas Anderson Keller ·
Nicu Sebe · Max Welling

Structured Representation Learning

From Homomorphisms and Disentanglement to Equivariance and Topography

Yue Song
Department of Computing and Mathematical Sciences
California Institute of Technology
Milan, Italy

Nicu Sebe
Department of Computer Science
and Information Engineering
University of Trento
Trento, Italy

Thomas Anderson Keller
Kempner Institute
Harvard University
Cambridge, MA, USA

Max Welling
Department of Machine Learning
University of Amsterdam
Amsterdam, The Netherlands

ISSN 2153-1056   ISSN 2153-1064 (electronic)
Synthesis Lectures on Computer Vision
ISBN 978-3-031-88110-7   ISBN 978-3-031-88111-4 (eBook)
https://doi.org/10.1007/978-3-031-88111-4

© The Editor(s) (if applicable) and The Author(s), under exclusive license to Springer Nature Switzerland AG 2026

This work is subject to copyright. All rights are solely and exclusively licensed by the Publisher, whether the whole or part of the material is concerned, specifically the rights of translation, reprinting, reuse of illustrations, recitation, broadcasting, reproduction on microfilms or in any other physical way, and transmission or information storage and retrieval, electronic adaptation, computer software, or by similar or dissimilar methodology now known or hereafter developed.
The use of general descriptive names, registered names, trademarks, service marks, etc. in this publication does not imply, even in the absence of a specific statement, that such names are exempt from the relevant protective laws and regulations and therefore free for general use.
The publisher, the authors and the editors are safe to assume that the advice and information in this book are believed to be true and accurate at the date of publication. Neither the publisher nor the authors or the editors give a warranty, expressed or implied, with respect to the material contained herein or for any errors or omissions that may have been made. The publisher remains neutral with regard to jurisdictional claims in published maps and institutional affiliations.

This Springer imprint is published by the registered company Springer Nature Switzerland AG
The registered company address is: Gewerbestrasse 11, 6330 Cham, Switzerland

If disposing of this product, please recycle the paper.

Preface I

The art of machine learning (ML) is to optimally combine inductive bias (a.k.a. priors) with data. With more data, we need to include less prior information and we can let the data speak. This is the modern "scaling" paradigm of LLMs and Foundation models with trillions of parameters, trained on hundreds of thousands of GPUs on the entirety of the internet. For these models, the architecture of choice are usually Transformers, precisely because they scale well.

Are Transformers the final answer? That seems unlikely. In this book, we ask ourselves if there are architectures based on better priors about the world but that also scale to internet-level models. These models will not only be able to learn from fewer data but also exhibit improved scaling laws, ideally with a steeper slope.

What are some interesting priors to build into deep architectures? In this book, we are inspired by both neuroscience and physics. Neuroscience has been ML's companion right from the start. Early architectures such as Rosenblatt's Perceptron were already inspired by biological neurons. It's only more recently that the two fields have gone their own separate ways. But given the huge gap in energy efficiency between artificial and biological neural networks, it may make sense to look at neuroscience again for inspiration.

In this book, we explore the possibility to use oscillators and traveling waves as a new computing paradigm, rather than static representations. Waves have the potential to collect and combine information from long distances, both on space and in time.

Another interesting prior is that in the world around us at the scale that we understand it (objects), things usually don't change very fast. This is of course different than the nanosecond fluctuation at the level of individual atoms. As deep models build coarse-grained representations in their deeper levels, it seems reasonable to enforce this slowness at the level of abstract (deep) representations.

Since our models often model things in our physical world, we can also contemplate whether the symmetries of that physical world should be represented in our representations. There is now a rich literature on symmetries, such as translational and rotational symmetries that are exactly enforced through (irreducible or regular) equivariant group representations. However, often we do not know the symmetries present in the data, or some regularities might not be described by groups, or we may not simply know the representations of certain groups. In all these cases we need to generalize the concept of equivariance, on which these hard-coded symmetries are based.

In this book, we consider homomorphisms between latent representations and the world as an appropriate signal to learn such approximate symmetries. That is to say, the dynamics of the latent representations should mirror the corresponding dynamics of the world. What dynamics, or set of transformation of our latent code, do we entertain when we try to learn homomorphisms between the world and our deep representations? Here we are inspired again by neuroscience and physics: we have modeled these representations as collections of interacting oscillators, or in the continuum limit, PDEs, that support wave-like solutions. In some sense, we can think of these representations as a fluid in which waves can develop to perform computations.

And this brings me to my final point. Due to availability of multi-electrode sensors, waves are now commonly detected in the brain, and neuroscience researchers are starting to ask what its computational benefits might be. Can waves transport and combine information in new ways that we have not yet discovered? This is an intriguing possibility about which I have no doubt we will hear a lot more over the course of the next decade.

I wish you an interesting journey as you travel through the chapters of this book and become inspired to think of new ways to build inductive biases into the next generation of ML models.

Amsterdam, The Netherlands   Max Welling
February 2025

Preface II

The field of machine learning stands at a critical juncture. While recent advances in deep learning have delivered remarkable breakthroughs across domains such as vision, language, and robotics, these successes often come at the cost of massive data requirements, computational inefficiency, a lack of interpretability, and poor generalization to novel scenarios. As machine learning systems are increasingly deployed in real-world applications, these challenges highlight the need for models that go beyond brute force learning and instead can learn more like humans—adaptively, efficiently, and intuitively.

This book, *Structured Representation Learning: From Homomorphisms and Disentanglement to Equivariance and Topography*, offers a timely exploration of how structured approaches can reshape the design and performance of machine learning systems. By embedding principles such as symmetry, topography, and compositionality directly into model architectures, structured representation learning provides a pathway to models that are more robust, efficient, and capable of generalization.

At its core, structured representation learning seeks to address fundamental questions: How can machine learning systems capture the inherent relationships within data, such as symmetries and invariances? How can models decompose complex phenomena into simpler, interpretable components? And how can we align computational representations with the physical and biological principles that govern the real world? This book explores these questions through key concepts such as:

- **Equivariance and Symmetry**: Learning approximately equivariant representations that respect the transformations of the data beyond group theory.
- **Disentanglement**: Designing latent representations that isolate meaningful factors of variation, serving as approximate learned equivariance.
- **Topographic Representations**: Drawing inspiration from biological systems to organize information spatially and temporally in ways that mimic biological neural networks.

- **Physical Priors**: Baking physical principles into machine learning systems to represent the physical relations in the real world.

Structured representation learning represents a paradigm shift. By incorporating the above beneficial inductive biases directly into learning systems, this approach opens the door to machine learning systems that are not only more efficient but also more interpretable and aligned with the complexities of the real world.

This book is written for researchers, practitioners, and students eager to explore the intersection of machine learning, computational neuroscience, and natural sciences. It provides a high-level perspective on the field's foundational ideas while delving into specific techniques and applications that demonstrate the power of structured representation learning. As the demands on machine learning systems continue to grow, structured representations offer a promising direction toward building models that can reason, adapt, and learn with greater data efficiency and generalization abilities. We invite readers to engage with the ideas in this book, explore the rich potential of structured representations, and join in shaping the future of machine intelligence.

Pasadena, CA, USA Yue Song
Cambridge, MA, USA Thomas Anderson Keller
November 2024

Contents

Part I Structured Representation Learning: History and State of the Art

1 Introduction 3
   1.1 Motivation: The Gap of Sample Efficiency and Generalization 3
   1.2 The Promise of Structured Representations 5
   1.3 The Way Forward: Naturally Inspired Learned Homomorphisms 7
   References 7

2 Background 9
   2.1 Structured Representation Learning 9
   2.2 Disentangled Representation Learning 10
   2.3 Equivariant Neural Networks 11
   2.4 Approximately Equivariant and Disentangled Representations 12
      2.4.1 Prior Work: Capsule Networks, Homeomorphic VAEs 12
      2.4.2 Biological and Physical Inductive Biases for Learning
            Equivariant Representations 12
   References 14

Part II Naturally Inspired Topographically Structured Representation Learning

3 Topographic Variational Autoencoders 21
   3.1 Introduction 21
   3.2 The Generative Model 22
      3.2.1 Topographic Generative Models 23
      3.2.2 The Product of Student's-t Model 23
      3.2.3 Constructing the Product of Student's-t Distribution 24
      3.2.4 Introducing Topography 24
      3.2.5 Capsules as Disjoint Topologies 25

		3.2.6	Temporal Coherence and Learned Equivariance	25
	3.3		Topographic VAE	27
	3.4		Experiments	27
		3.4.1	Evaluation Methods	28
		3.4.2	Topographic VAE Without Temporal Coherence	29
		3.4.3	Learning Equivariant Capsules	29
	3.5		Future Work and Limitations	32
	3.6		Conclusion	32
	3.7		Experiment Details	33
		3.7.1	Optimizer Parameters	33
		3.7.2	Initalization	33
		3.7.3	Model Architectures	33
		3.7.4	Choices of $\mathbf{W}$, $\mathbf{W}_\delta$, and L	34
		3.7.5	Hyperparameter Selection	35
		3.7.6	MNIST Transformations	35
		3.7.7	dSprites Transformations	35
		3.7.8	Capsule Correlation Metric (CapCorr)	36
		3.7.9	Definition of Roll for Capsules	37
	3.8		Extended Results	37
		3.8.1	Extended Tables 3.1 and 3.2	38
		3.8.2	Impact of $\mathbf{W}_\delta$	39
		3.8.3	Generalization to Combined Transformations at Test Time	39
	3.9		Proposed Model Extensions	41
		3.9.1	Extensions to Roll and CapCorr	41
		3.9.2	Non-cyclic Capsules	43
		3.9.3	Multi-dimensional Temporally Coherent Capsules	44
		3.9.4	Causal Temporal Coherence	44
	3.10		Capsule Traversals	45
	References			47
4	**Neural Wave Machines**			49
	4.1		Introduction	49
		4.1.1	Traveling Waves in Neuroscience	51
		4.1.2	Computational Models of Traveling Waves	52
	4.2		Neural Wave Machines	52
		4.2.1	Coupled Oscillatory Recurrent Neural Networks	52
		4.2.2	Local Connectivity	53
	4.3		Experiments	54
		4.3.1	Methods	54
		4.3.2	Datasets	55
		4.3.3	Measuring Spatiotemporal Structure	56
		4.3.4	Topographic Orientation Selectivity	56

	4.3.5	General Topographic Organization	57
	4.3.6	Instantaneous Phase and Velocity	57
	4.3.7	Controlled Generation with Induced Traveling Waves	58
	4.3.8	Computational Implications of Structure	60
	4.3.9	An Inductive Bias for Simple Physical Dynamics	60
	4.3.10	Efficiency	60
4.4	Discussion		62
	4.4.1	Related Work	62
	4.4.2	Limitations	62
	4.4.3	Conclusion	63
4.5	Experiment Details		63
	4.5.1	Sequence Classification	64
	4.5.2	Rotating MNIST and Sine Waves	64
	4.5.3	Hamiltonian Dynamics Suite	66
	4.5.4	Hardware Details	67
4.6	Analytical Treatment of Neural Wave Machines		67
	4.6.1	Bounds on Hidden State Energy	68
	4.6.2	Sensitivity to Inputs	68
	4.6.3	Bounds on Hidden State Gradient	68
	4.6.4	Assumptions	69
4.7	Extended Results		69
	4.7.1	Impact of Δt Parameter	69
	4.7.2	Additional Efficient Sequence Modeling Results	70
	4.7.3	Additional Hamiltonian Dynamics Results	70
	4.7.4	On Modeling Chaotic Dynamics	71
	4.7.5	On the Formation of Orientation Maps	72
	4.7.6	Full Rotating MNIST Topographic Organization	73
	4.7.7	Visualizing Traveling Waves on MNIST	74
References			77

Part III Learned Homomorphisms and Disentangled Representations

5 Latent Traversal as Potential Flows 83
    5.1   Motivations 83
           5.1.1   Traveling Waves in Neuroscience 83
           5.1.2   Fluid Mechanics as Optimal Transport 83
    5.2   Learned Potential Flows for Traversal 85
           5.2.1   Learning the Potential PDEs 85
           5.2.2   Integration with Generative Models 86
    5.3   Experiments 88
           5.3.1   Evaluation Methods 88
           5.3.2   Results with Pre-trained Networks 89

		5.3.3 Results with Pre-trained VAEs	90
		5.3.4 Results with Networks Trained from Scratch	91
	5.4	Discussions	93
		5.4.1 Flow Path Properties	93
		5.4.2 Limitations and Future Extensions	95
	References		96
6	**Flow Factorized Representation Learning**		**99**
	6.1	Factorized Representation Learning	99
		6.1.1 Learned Equivariance	99
		6.1.2 Disentanglement	99
	6.2	The Generative Model	100
		6.2.1 Flow Factorized Sequence Distributions	100
		6.2.2 Prior Time Evolution	101
	6.3	Flow Factorized Variational Autoencoders	102
		6.3.1 Inference with Observed Transformation Categories	102
		6.3.2 Inference with Unknown Transformation Categories	103
		6.3.3 Posterior Time Evolution	104
		6.3.4 Optimal Transport for Posterior Flow	104
	6.4	Experiments	106
		6.4.1 Evaluation Methods	106
		6.4.2 Learning Equivariant Latent Flows	107
		6.4.3 Results on Complex Real-World Datasets	110
	6.5	Discussions	110
		6.5.1 Extrapolation to Switching/Superposing Transformation	110
		6.5.2 Equivariance Generalization to New Data	113
	References		114
7	**Unsupervised Factorized Representation Learning Based on Sparse Transformation Analysis**		**117**
	7.1	Motivations	117
		7.1.1 Sparsity in Natural Videos	117
		7.1.2 Helmholtz Decomposition	118
	7.2	The Generative Model	118
		7.2.1 Factorized Sequence Distributions	118
		7.2.2 Spike and Slab Priors	119
		7.2.3 Prior Time Evolution	121
	7.3	Helmholtz Flow Variational Autoencoders	121
		7.3.1 Helmholtz Decomposed Latent Flows	121
		7.3.2 Evidence Lower Bound and Inference	122
		7.3.3 Posterior Time Evolution	123
	7.4	Experiments	124

		7.4.1	Evaluation Methods	124
		7.4.2	Learning Composable Equivariant Latent Flows	125
		7.4.3	Analysis of Real-World Videos	129
	7.5	Discussions		134
		7.5.1	Switchability and Composability	134
		7.5.2	Handling Periodic Transformations	134
		7.5.3	Learning Separate Controls	135
	References			136
8	**Conclusion**			139
	8.1	Naturally Inspired Structured Representations		139
	8.2	Learned Homomorphisms and Disentangled Representations		140
	Reference			140

Acronyms

AI	Artificial Intelligence
AR	AutoRegressive Model
ELBO	Evidence Lower BOund
GAN	Generative Adversarial Network
GPU	Graphics Processing Unit
HJE	Hamilton-Jacobi Equation
ICA	Independent Component Analysis
LLM	Large Language Models
LN	Layer Normalization
LSTM	Long Short-Term Memory
MLP	Multi-layer Perceptron
ODE	Ordinary Differential Equations
OoD	Out-of-Distribution
OT	Optimal Transport
PDE	Partial Differential Equation
PINN	Physics-Informed Neural Network
RNN	Recurrent Neural Network
SE(N)	Special Euclidean Group
SFA	Slow Feature Analysis
SO(N)	Special Orthogonal Group
SOM	Self-Organizing Map
VAE	Variational Auto-Encoder

List of Figures

Fig. 1.1   Samples from text-to-image generation program DALLE-2. The prompts are: (top) "a teddy bear on the moon", (middle) "blue cube on top of a red cube", & (bottom) "a banana holding a monkey". We see that while the model can generate incredibly realistic images of relatively novel objects, it often fails to understand basic relations between objects 4

Fig. 3.1   Overview of the Topographic VAE with shifting temporal coherence. The combined color/rotation transformation in input space τ_g becomes encoded as a Roll within the capsule dimension. The model is thus able decode unseen sequence elements by encoding a partial sequence and Rolling activations within the capsules. We see this resembles a commutative diagram 22

Fig. 3.2   An example of a neighborhood structure which induces disjoint topologies (A.K.A. capsules). Lines between variables T_i indicate that sharing of U_i, and thus correlation 26

Fig. 3.3   Maximum activating images for a topographic VAE trained with a 2D torus topography on MNIST 30

Fig. 3.4   Capsule Traversals for TVAE models on dSprites and MNIST. The top rows show the encoded sequences (with greyed-out images held-out), and the bottom rows show the images generated by decoding sequentially Rolled copies of the initial activation $\mathbf{t}_0$ (indicated by a grey border) 30

Fig. 3.5   Capsule Traversals for MNIST TVAE $L = \frac{13}{36}S$, trained on individual transformations in isolation, and tested on combined color and rotation transformations. Top row shows the input sequence, middle row shows the direct reconstruction $\{g_\theta(\mathbf{t}_l)\}_l$, and bottom row shows the capsule traversal $\{g_\theta(\text{Roll}_l[\mathbf{t}_0])\}_l$ 39

Fig. 3.6   MNIST TVAE $L = \frac{1}{2}S$, $K = 3$ 40

Fig. 3.7   MNIST TVAE $L = \frac{13}{36}S$, $K = 3$ 40

Fig. 3.8   MNIST TVAE $L = \frac{5}{36}S$, $K = 3$. We see with values of $L < \frac{1}{3}S$ the transformations decoded through the capsule roll are only partially coherent with the input sequence 41

Fig. 3.9   MNIST TVAE $L = \frac{5}{36}S$, $K = 9$ 41

Fig. 3.10 MNIST TVAE $L = 0$, $K = 3$. We see for sufficiently small values of K, the TVAE can reach a degenerate solution where topographic organization is almost entirely lost 42

Fig. 3.11 MNIST TVAE $L = 0$, $K = 9$ 42

Fig. 3.12 MNIST TVAE $L = 0$, $K = 18$. We see when K is equal to the capsule size (making the model analogous to ISA), the model learns an invariant capsule representation – meaning Rolling a capsule activation produces no significant transformation in the observation space 43

Fig. 3.13 MNIST VAE $L = 0$, $K = 1$. We see images generated through capsule traversal with the baseline VAE appear entirely random, as expected due to the non-topographic nature of the VAE's latent space 45

Fig. 3.14 Combined Color & Rotation MNIST TVAE $L = \frac{13}{36}S$, $K = 3$. We see these generated sequences are slightly more accurate than those in Fig. 3.5. This is to be expected since the model in this figure is trained explicitly on combinations of transformations, whereas the model in Fig. 3.5 was trained on transformations in isolation, and tested on combinations to explore its generalization 46

Fig. 3.15 Combined Color & Perspective MNIST TVAE $L = \frac{13}{36}S$, $K = 3$. We see the TVAE is able to additionally learn combinations of complex transformations (like out-of-plane rotation) without any changes to the training procedure other than a change of dataset 46

List of Figures   xix

Fig. 4.1 Overview of the Neural Wave Machine. The input sequence $\mathbf{u}$ is encoded with f_θ to act as a driving term in the hidden state $\mathbf{x}$ which is modeled temporally ($\ddot{\mathbf{x}}$) as a network of locally coupled oscillators. The network is then trained to reconstruct the input sequence: $\hat{\mathbf{u}} = g_\theta(\mathbf{x})$. The yellow arrows track a traveling wavefront over time 50

Fig. 4.2 Plot of different datasets used in this work (top) and the associated learned hidden state dynamics (bottom). We see the NWM learns different spatiotemporal structure for each dataset, and no structure when trained on random noise (a). Additional videos of dynamics, and code for experiments, can be found at: github.com/akandykeller/NeuralWaveMachines 55

Fig. 4.3 (Left) Plot of orientation selectivity of each NWM hidden neuron $\mathbf{x}$ after training on simple sine waves. (Right) Plot of the maximum activating image for a subset of NWM hidden neurons after training on the rotating MNIST dataset (See Sect. 4.7.6 for full). We see the NWM learns smooth spatial topographic structure tailored to the input dataset 57

Fig. 4.4 (Left) Plot of hidden state $\mathbf{x}$ (top), generalized phase ϕ (mid), and estimated wave velocity $-\nabla\phi$ (bot) over the course of a transformation sequence $T = 0$ to 3. A small gold star moves along with a wave front, relative to a stationary grey triangle, both added to help track the approximate peak of a traveling wave in the hidden state. (Right) Estimated wave velocity before and after training 58

Fig. 4.5 Visualization of controlled generation with induced traveling waves. An input sequence from $\mathbf{u}_0$ to $\mathbf{u}_T$ (left) gets encoded to a hidden state $\mathbf{x}_T$. We then induce a traveling wave in the opposite direction of the estimated instantaneous velocity and observe we can decode back to the original input $\hat{\mathbf{u}}_0$ (highlighted yellow, right). Furthermore, we see by continuing the wave, we can continue the transformation past the bounds of the input sequence (highlighted pink, right) 59

Fig. 4.6 Orientation selectivity (left) and instantaneous phase at a random sequence element (right) for a model trained on the sine waves dataset. We see that the phase synchrony across the neurons is roughly in alignment with the orientation selectivity, supporting the hypothesis that this is one of the primary mechanisms for topographic organization in the NWM 72

Fig. 4.7 Orientation selectivity maps as a function of training dataset wavelength (λ^{train}), and kernel size (size($\mathbf{w}_z$)) 73

Fig. 4.8	Depiction of the maximum activating image for the full set of neurons in the NWM when training on Rotated MNIST. The subset depicted in Fig. 4.3 is highlighted in yellow. We see that topographic organization is widespread and roughly continuous throughout the hidden state	74
Fig. 4.9	Additional hidden state visualizations for the model in Fig. 4.4. Reconstructions (Top), Hidden state (middle) and generalized phase (bottom), for the final 18 timesteps of the test sequence	75
Fig. 4.10	Visualization of the hidden state and phase for three models identical to those in Fig. 4.4, but with different random initalizations. We see that the models learn different wavelengths and velocities depending on their initialization	75
Fig. 4.11	Additional visualizations of reconstructions from induced wave activity in the hidden state of the 1D NWM as depicted in Fig. 4.5. We show a set of random input sequences (top), the original model reconstruction (middle), and images generated by sequentially propagating the initial state backwards by an induced wave and decoding at each step (bottom). We see that, as in the main text, the assumed wave velocity of $v = 1$ is slightly faster than the actual velocity, and thus the reconstructed transformations are slightly faster than the input transformations. Because of this, we also observe that for certain examples, the induced wave reconstructions lose consistency with the input after the first period. This appears to imply that both the initial location of the wave activity matters in addition to its wave properties, and thus our model has learned to only propagate waves over parts of the feature space to optimize the capacity of the hidden state for this dataset. Finally, we observe that the induced transformations occur in reverse order due to the fact that our induced waves propagate in the reverse direction to those naturally exhibited for training examples, effectively propagating backwards in time	76
Fig. 5.1	Overview of our learned PDEs for latent traversal in two different experimental settings	85

Fig. 5.2	Exemplary traversal paths (potential PDEs for our method) and the corresponding interpolation images with SNGAN and BigGAN. Since the paths of WarpedSpace are of very limited non-linearity that is hard to perceive, we amplify the non-linear part in the sub-figure inside the figure as follows: for a traversal path y of WarpedSpace, we decompose it into $y = y_{LN} + y_{NLN}$ where y_{LN} denotes the linear part and y_{NLN} is the non-linear counterpart. Then the non-linearity part is amplified by $y = y_{LN} + 200 \cdot y_{NLN}$	89
Fig. 5.3	Traversal trajectories (potential PDEs for our method) and the associated interpolation images of the exemplary four attributes with StyleGAN2. The non-linearity of WarpedSpace paths is amplified in the same way as done in SNGAN and BigGAN	91
Fig. 5.4	Exemplary semantic attributes and the corresponding traversal trajectories with VAEs trained on MNIST and dSprites	92
Fig. 5.5	Exemplary traversal results when our method is integrated into the VAE training process. For MNIST, the exhibited transformations are scaling, rotation, and coloring changes from top to bottom. For Despites, the corresponding transformations are y-axis position, scaling, and shape changes from top to bottom	93
Fig. 5.6	Common shapes of potential PDEs in our experiments	94
Fig. 5.7	Unambiguity of our potential PDEs and the corresponding discovered semantics: the shape of trajectory and the image attribute of a traversal path are consistent to different samples	95
Fig. 6.1	Illustration of our flow factorized representation learning: at each point in the latent space we have a distinct set of tangent directions ∇u^k which define different transformations we would like to model in the image space. For each path, the latent sample evolves to the target on the potential landscape following dynamic optimal transport	100
Fig. 6.2	Depiction of our model in plate notation. (Left) Supervised, (Right) Weakly-supervised. White nodes denote latent variables, shaded nodes denote observed variables, solid lines denote the generative model while dashed lines denote the approximate posterior. We see, as in a standard variational autoencoder, our model approximates the initial one-step posterior $p(z_0|x_0)$, but additionally approximates the conditional transition distribution $p(z_t|z_{t-1}, k)$ through optimal transport over an approximate potential landscape	101

Fig. 6.3	Exemplary latent evolution results of Scaling, Rotation, and Coloring on MNIST [6]. The top two rows are based on the supervised experiment, while the images of the bottom row are taken from the weakly-supervised setting of our experiment	107
Fig. 6.4	Exemplary latent flow results on Shapes3D [7]. The transformations from top to bottom are Floor Hue, Wall Hue, Object Hue, and Scale, respectively. The images of the top row are from the supervised experiment, while the bottom row is based on the weakly-supervised experiment	108
Fig. 6.5	Qualitative comparison of our method against TVAE and PoFlow on Falcol3D and Isaac3D	113
Fig. 6.6	Exemplary visualization of switching flow fields during the latent sample evolution	113
Fig. 6.7	Examples of combining flow fields simultaneously during latent evolution	114
Fig. 6.8	Equivariance generalization to unseen OoD input data. Here the model is trained on MNIST [6] but the latent flow is tested on dSprites [16]	114
Fig. 7.1	Our model across N sequences in plate notation (Left) and a detailed version with decomposed spike and slab components (Right). White nodes denote latent variables, shaded nodes denote observed variables, solid lines denote the generative model, and dashed lines denote the approximate posterior. Different from the spike component y_t, the slab variable $\tilde{g}_t$ is independent across timesteps	119
Fig. 7.2	Exemplary sequences generated by our spike prior	120
Fig. 7.3	Traversals using individual learned flows $k=\{0, 1, 2\}$ from left to right with speeds $g_t=\{\frac{1}{2}, 1, 2\}$ from top to bottom	125
Fig. 7.4	Traversals using each individual learned flow field on Shapes3D [13]. In the bracket, we indicate the transformation which the traversal results look most like. Each latent flow has separate samples per row transforming from left to right	126
Fig. 7.5	Traversals using learned flows with different speeds $g_t = \{\frac{1}{2}, 1\}$ on Shapes3D	126
Fig. 7.6	Traversals using each individual learned flow field on Falcol3D [14]. In the bracket, we indicate the transformation which the traversal results look most like. Each latent flow has separate samples per row transforming from left to right. The bottom row displays the traversal result generated by the 6'th latent flow field	129

Fig. 7.7	Traversals using each individual learned flow field on Issac3D [14]. In the bracket, we indicate the transformation which the traversal results look most like. Each latent flow has separate samples per row transforming from left to right	130
Fig. 7.8	Traversal results of learned latent flows on CalMS [16]. For each latent flow, we display two exemplary sequences, and the flow transforms the image from left to right	133
Fig. 7.9	Exemplary comparisons of the ground truth image sequences and reconstruction results. For each sequence, we start with reconstructing the initial frame and use the spike component and latent flow fields to generate the rest frames	133
Fig. 7.10	Traversals results of learned flow fields on downsampled segmentation masks of Cityscape [15]. Each latent flow transforms the image from left to right	134
Fig. 7.11	Traversal results of switching latent flows	134
Fig. 7.12	Traversal results of combining latent flows	135
Fig. 7.13	Traversal results using different types of vector fields	135

List of Tables

Table 3.1	Log Likelihood and Equivariance Error on MNIST for different settings of temporal coherence length L relative to sequence length S. Mean ± std. over 3 random initalizations	31
Table 3.2	Equivariance error ($\mathcal{E}_{eq}\downarrow$) and correlation of observed capsule roll with ground truth factor shift (CapCorr ↑) for the dSprites dataset. Mean ± standard deviation over 3 random initalizations	31
Table 3.3	Log Likelihood and Equivariance Error on MNIST for all models tested. Mean ± std. over 3 random initalizations	37
Table 3.4	Equivariance error and CapCorr for all models tested on the dSprites dataset. Mean ± standard deviation over 3 random initalizations.	38
Table 3.5	Impact of $\mathbf{W}\,\delta$ (i.e. K) on MNIST performance	38
Table 4.1	Forward extrapolation mean squared reconstruction error on the Hamiltonian Dynamics Benchmark held-out test set (displayed in units of 1×10^{-8}). We see, in alignment with intuition, the 1 and 2-dimensional Neural Wave Machines (NWM 1D & 2D) perform best on simple physically realistic dynamics such as the spring, pendulum, and two body problem. The globally coupled coRNN performs best on the smooth, but non-physical, matching pennies task, while the maximally flexible Neural ODE performs best on the highly complex and chaotic double pendulum task	61
Table 4.2	Test accuracy on supervised sequence benchmarks. All results are mean ± std. over 3 random initalizations	61
Table 4.3	Test accuracy on the sMNIST dataset for a range of Δt values	69

Table 4.4	Test accuracy on additional sequence modeling benchmarks including the long-sequence Addition task from [53], and the IMDB sentiment classification task. All results are mean ± std. over 3 random initalizations. We see similar results to those shown in Table 4.2, the NWM achieves comparable performance while requiring significantly fewer parameters	70
Table 4.5	Valid Prediction Time 'VPT' (± std.) on the Hamiltonian Dynamics Benchmark. We highlight in bold results which fall within one standard deviation of the best performing model. We see that the VPT metric has large standard deviation owing to the reliance on an arbitrary threshold of image-space similarity, however the NWM models still perform favorably compared with existing state of the art	71
Table 4.6	Test Mean Squared Error of an LSTM and NWM when forecasting the Lotentz '96 attractor. We see that the NWM performs better in the non-chaotic regime ($F = 0.9$), while in chaotic regime ($F = 8$) the LSTM performs significantly better	71
Table 5.1	Comparison of the VP scores (%) with different GANs averaged over 3 random runs	90
Table 5.2	The l_1 normalized attribute correlations of our method (*top*), WarpedSpace (*middle*), and SeFa (*bottom*) based on 50 samples. The second highest correlation is also highlighted if the best value in the row is not on the diagonal	91
Table 5.3	Comparison of the VP scores (%) with pre-trained VAEs averaged over 3 random runs	92
Table 5.4	The log-likelihood $\log p_\theta(x)$ evaluated over the dataset	92
Table 5.5	Equivariance error on MNIST	93
Table 6.1	Equivariance error ε_k and log-likelihood $\log p(x_t)$ on MNIST [6]	108
Table 6.2	Equivariance error ε_k and log-likelihood $\log p(x_t)$ on Shapes3D [7]	109
Table 6.3	VP Scores (%) on MNIST	110
Table 6.4	VP scores (%) on Shapes3D	110
Table 6.5	Equivariance error (↓) on Falcol3D [8]	111
Table 6.6	Equivariance error (↓) on Isaac3D [8]	112
Table 7.1	Equivariance error ε_k and average log-likelihood $\log p(xt)$ on MNIST [9]	127
Table 7.2	VP Scores (%) on MNIST	127
Table 7.3	VP Scores (%) on Shapes3D	127
Table 7.4	Equivariance error ε_k and average log-likelihood $\log p(x\ t)$ on Shapes3D [13]	128

Table 7.5	Equivariance error ε_k of composite transformations. For both baselines, we linearly combine their latent flows	129
Table 7.6	Equivariance error ε_k on Falcor3D	131
Table 7.7	Equivariance error ε_k on Issac3D	132
Table 7.8	Behavior classification results on CalMS [16]	133
Table 7.9	The learned association of different vector fields for each transformation on MNIST	136

Part I
Structured Representation Learning: History and State of the Art

Introduction 1

1.1 Motivation: The Gap of Sample Efficiency and Generalization

Modern artificial neural networks are known to require an inordinate amount of training data in order to reach the state of the art performance we have now become accustomed to. Consider, for example, the groundbreaking Go-playing program, AlphaGo. Its human opponent in the highly publicized match, Lee Sedol, is estimated to have played on the order of 100,000 games throughout his training lifetime, ultimately achieving the highest possible rank of 9th dan. Comparatively, Alpha Go is believed to have learned from nearly 100 million or more games altogether [1]. To understand the scale of this difference, if each letter on this page were to represent a single game, a human player would require around 30 pages of text. For AlphaGo, the text would fill nearly 30,000 pages, or more than 150 times the size of this book. In a more controlled setting, discrepancies between the speeds of human and artificial learning efficiency have been studied experimentally by scientists such as Brendan Lake et al. [1]. In their work, these authors report that humans are able to learn to effectively play a suite of Atari games in just two hours, reaching a level of performance that took modern deep reinforcement learning agents an equivalent of 924 h of game time to learn [2].

However, it is well known that reinforcement learning agents are not the only algorithms which require orders of magnitude more data than humans to perform comparably. Large language models (LLMs) are trained on nearly the entirety of the internet in order to be able to answer natural questions reasonably. LaMBDA, an early LLM from Google, is stated to have been trained on 1.56 trillion tokens extracted from public dialog and text [3]. As a point of comparison, it is estimated the average English non-fiction reading speed is roughly 240 words per minute [4], meaning that it would take the average human 18,550 years of reading non-stop, 16 h per day, to ingest the same amount of data. Even more recent models such as GPT-4 are rumored to have been trained on near 13 trillion tokens, equating to more than 150,000 years of human reading [5]. Although these systems arguably also possess significantly more information internally than any living human being, it is clear that they do not behave in nearly the same manner as a someone who has managed to survive long enough to read an equivalent amount of text. In this book, we believe that this kind of behavioral mismatch is a telltale sign of the second greatest limitation of modern artificial intelligence, namely how these systems generalize to new situations from their necessarily finite training set.

It is clear from even cursory uses of modern language models or image generative models that these models do not generalize in any manner that resembles how humans generalize. For example, although modern text-to-image generation programs are able to generate highly photo-realistic images which appear to largely match their text-based prompts, their ability to generate slightly unorthodox images is still surprisingly lacking. Consider the examples shown in Fig. 1.1 from the state of the art generative model known as DALL-E 2 [6]. Although the images are highly photo-realistic, contain the mentioned objects, and are far better than anything most humans could draw or even create digitally, they appear to have fundamentally

Fig. 1.1 Samples from text-to-image generation program DALLE-2. The prompts are: (top) "a teddy bear on the moon", (middle) "blue cube on top of a red cube", & (bottom) "a banana holding a monkey". We see that while the model can generate incredibly realistic images of relatively novel objects, it often fails to understand basic relations between objects

misunderstood core elements of the text prompt, such as relatively simple relations between objects. Consider a small child as a point of comparison. Although it is clear that a child would never be able to draw the detailed lighting, shading, and texture of these images at a similar level of complexity, a child would likely be able to at least correctly draw a stick-figure banana holding a stick-figure monkey without getting permanently confused. This counter-intuitive performance reminds us of Moravec's paradox [7]—when it comes to artificial intelligence (AI) research: *"the hard problems are easy and the easy problems are hard"* [8]. We've developed machines that appear to surpass Leonardo da Vinci's level of control over lighting and perspective yet simultaneously fail at stacking blocks in the same manner as a toddler.

What is it about these systems that makes their generalization so counter-intuitive to us? What makes some tasks so easy for us, and so challenging to reproduce in silicon? In this book, we will argue that the answer, and foundation of these limitations of modern artificial intelligence comes from the lack of an inductive bias towards structure in neural representations.

1.2 The Promise of Structured Representations

One of the biggest lessons learned from the success of the deep learning revolution is the importance of 'good' representations. Specifically, the success of modern deep neural networks can largely be attributed to their ability to learn useful features directly from the data itself, rather than having these features hand-engineered by model developers. Yet, as these models have developed in sophistication, it has become increasingly clear that their ultimate flexibility is truly a double-edged sword.

Concretely, the fact that deep neural networks very easily 'overfit' to their data distribution, and struggle to generalize strongly out of domain, as exemplified by the image generation example in the previous subsection, can largely be attributed to the tendency of networks to learn features (or representations) tuned to their individual datasets rather than with respect to the data generating distribution as a whole. This limitation has sparked interest in integrating specific structural inductive biases into learned representations such that they are forced to learn more generalizable features. To date, one of the most successful methods for doing this has been through the introduction of what is known as 'equivariance'.

In a general sense, equivariance can be viewed as an inductive bias towards representations with geometric group structure (i.e. symmetries). Slightly more precisely, equivariance of a function can be understood to mean that for a given set of input transformations of interest, there is a corresponding known and well-behaved transformation of the function's output in the output space. In the language of group theory and representation theory, this is commonly written as $f(\tau_g[x]) = \Gamma_g[f(x)]$ for a function f, a transformation g, and the associated input and output representations of that transformation τ_g and Γ_g respectively. In the case of convolution, the operation of translating (i.e. shifting) the input can be seen as equivalent to translating the output of the network in feature space—in other words, the convolution layer is equivariant with respect to translation.

Another intuitive way to think of equivariance, which will come up multiple times in this book, is that it implies the function (or neural network) can be seen to commute with the transformation operator. In other words, it does not matter if one first transforms the input and then passes it through the function ($f(\tau_g x)$), or if one instead passes the un-transformed input through the function and then applies the transformation to the output ($\Gamma_g[f(x)]$), the result will be the same. We can draw this in the form of a diagram as follows:

$$\begin{array}{ccc} \mathcal{X} & \xrightarrow{\tau_g} & \mathcal{X}' \\ f\downarrow & & \downarrow f \\ \mathcal{Z} & \xrightarrow{\Gamma_g} & \mathcal{Z}' \end{array}$$

We see that, starting from $\mathcal{X}$, no matter whether we follow the upper path or the lower path, we arrive at the same point $\mathcal{Z}'$. As we will see throughout this work, many of our figures will resemble this type of commutative diagram to show that our network does indeed commute with the observed transformations.

In the years since the convolutional layer's widespread adoption, significant work has gone on to generalize the set of transformations to which networks can be made equivariant. These supported transformation groups now include rotations and mirroring [9], scaling [10], and ultimately any continuous compact Lie group [11]. In the simplest setting of group equivariant convolutional neural networks [9], rather than having filters defined over just space (the translation group), the filters now must be defined over the full group. In other words, we must have a transformed copy of each of the filters for each of the group elements g (i.e. each rotation angle or scale). Similarly, the output of the network will now have a separate output value for each element of the group, given by applying the transformed filter to the input. When a transformation is then observed to be applied to the input, one sees that the corresponding transformed filter will be selectively responsive. In this way, the output of the network is *structured* with respect to the given group transformation—the designer of the network knows how the output will transform for a given transformation of the input. These types of networks with structured representations have been demonstrated both empirically [9, 12–14] and theoretically [15–17] to improve data efficiency and generalization performance when the transformation groups they incorporate are reflected in the data they are aiming to model.

Despite the tremendous success of equivariance as a guiding principle for neural network architectural design, it is still not known how to construct networks with equivariance with respect to many natural transformations, such as lighting or perspective shift, due to their complex non-group structure. How then might natural systems handle the dramatic changes in lighting from day to day without getting confused? This will be one of the motivating questions for the work presented in this book. Precisely, in this work we will argue that the natural structure that we observe in the brain (i.e. spatial and spatio-temporal organization) may facilitate the learning of such approximately equivariant architectures with respect to these more complex natural transformations.

1.3 The Way Forward: Naturally Inspired Learned Homomorphisms

As stated previously, to build an equivariant neural network, analytic methods require that the desired transformation conforms to the mathematical properties of a group, and that the corresponding representation theory for that group is known. In the case of common natural transformations, such as changes of lighting, changes of facial pose, or even simple occlusions, these assumptions do not hold.

To make progress, we consider if it is possible to mimic the essential elements of equivariance in order to generalize the benefits of this type of structure to a broader class of transformations. In the chapters of this book we will introduce a number of possible such generalizations which range from optimal transport flows to topographic organization. In each of these cases, portions of these methods are indeed inspired by physics and biology, aiming to replicate the inductive biases which we know much exist in natural learning systems, enabling their efficiency and generalization.

One overarching theme between the chapters presented in this book is that of learned homomorphisms. At a high level, the concept of a homomorphism is a map that preserves the algebraic structure of the input space in the output space. Group equivariant neural networks are group-homomorphisms: they preserve the structure of the transformation group from the input to the output. If we do not know the representation theory of the transformations we are interested in, however, perhaps we can enforce an explicit transformation in latent space, and learn the encoders. The main requirement being that the transformations we enforce in the latent space must respect the abstract structure of the input transformations. Alternatively, if we have a pre-trained encoder, can we learn the corresponding latent transformations after training? Exploring these ideas thoroughly will encompass the remainder of this book and will form the basis of what we call learned homomorphisms.

References

1. Brenden M. Lake, Tomer D. Ullman, Joshua B. Tenenbaum, and Samuel J. Gershman. Building machines that learn and think like people. *Behavioral and Brain Sciences*, 40:e253, 2017.
2. Tom Schaul, John Quan, Ioannis Antonoglou, and David Silver. Prioritized experience replay, 2016.
3. Romal Thoppilan, Daniel De Freitas, Jamie Hall, Noam Shazeer, Apoorv Kulshreshtha, Heng-Tze Cheng, Alicia Jin, Taylor Bos, Leslie Baker, Yu Du, YaGuang Li, Hongrae Lee, Huaixiu Steven Zheng, Amin Ghafouri, Marcelo Menegali, Yanping Huang, Maxim Krikun, Dmitry Lepikhin, James Qin, Dehao Chen, Yuanzhong Xu, Zhifeng Chen, Adam Roberts, Maarten Bosma, Vincent Zhao, Yanqi Zhou, Chung-Ching Chang, Igor Krivokon, Will Rusch,

Marc Pickett, Pranesh Srinivasan, Laichee Man, Kathleen Meier-Hellstern, Meredith Ringel Morris, Tulsee Doshi, Renelito Delos Santos, Toju Duke, Johnny Soraker, Ben Zevenbergen, Vinodkumar Prabhakaran, Mark Diaz, Ben Hutchinson, Kristen Olson, Alejandra Molina, Erin Hoffman-John, Josh Lee, Lora Aroyo, Ravi Rajakumar, Alena Butryna, Matthew Lamm, Viktoriya Kuzmina, Joe Fenton, Aaron Cohen, Rachel Bernstein, Ray Kurzweil, Blaise Aguera-Arcas, Claire Cui, Marian Croak, Ed Chi, and Quoc Le. Lamda: Language models for dialog applications, 2022.
4. Marc Brysbaert. How many words do we read per minute? a review and meta-analysis of reading rate. *Journal of Memory and Language*, 109:104047, 2019.
5. Maximilian Schreiner. Gpt-4 architecture, datasets, costs and more leaked, Jul 2023.
6. Aditya Ramesh, Prafulla Dhariwal, Alex Nichol, Casey Chu, and Mark Chen. Hierarchical text-conditional image generation with clip latents, 2022.
7. Hans P Moravec. *Mind children*. Harvard University Press, London, England, July 1990.
8. Steven Pinker. *The language instinct*. HarperCollins, New York, NY, 2007.
9. Taco Cohen and Max Welling. Group equivariant convolutional networks. In *ICML*, 2016.
10. Daniel Worrall and Max Welling. Deep scale-spaces: Equivariance over scale. In H. Wallach, H. Larochelle, A. Beygelzimer, F. d'Alché-Buc, E. Fox, and R. Garnett, editors, *Advances in Neural Information Processing Systems*, volume 32. Curran Associates, Inc., 2019.
11. Marc Finzi, Samuel Stanton, Pavel Izmailov, and Andrew Gordon Wilson. Generalizing convolutional neural networks for equivariance to lie groups on arbitrary continuous data. In *ICML*. PMLR, 2020.
12. Daniel E Worrall, Stephan J Garbin, Daniyar Turmukhambetov, and Gabriel J Brostow. Harmonic networks: Deep translation and rotation equivariance. In *CVPR*, 2017.
13. Bastiaan S. Veeling, Jasper Linmans, Jim Winkens, Taco Cohen, and Max Welling. Rotation equivariant cnns for digital pathology. *CoRR*, abs/1806.03962, 2018.
14. Elise van der Pol, Daniel E. Worrall, Herke van Hoof, Frans A. Oliehoek, and Max Welling. MDP homomorphic networks: Group symmetries in reinforcement learning. *CoRR*, abs/2006.16908, 2020.
15. Bryn Elesedy and Sheheryar Zaidi. Provably strict generalisation benefit for equivariant models. In Marina Meila and Tong Zhang, editors, *Proceedings of the 38th International Conference on Machine Learning*, volume 139 of *Proceedings of Machine Learning Research*, pages 2959–2969. PMLR, 18–24 Jul 2021.
16. Matthew Farrell, Blake Bordelon, Shubhendu Trivedi, and Cengiz Pehlevan. Capacity of group-invariant linear readouts from equivariant representations: How many objects can be linearly classified under all possible views?, 2021.
17. Blake Bordelon and Cengiz Pehlevan. Population codes enable learning from few examples by shaping inductive bias. *eLife*, 11:e78606, dec 2022.

Background 2

2.1 Structured Representation Learning

The resounding success of deep learning in the last decade has largely been attributed to the ability of deep neural networks to learn valuable internal representations directly from data. Such representations are now at the forefront of many of today's most advanced technologies, allowing for the extraction of abstract semantics from high dimensional data, and enabling previously unimaginable technologies such as automatic image inpainting and apparent natural language understanding. Despite the impressive affordances of representation learning, the generalization abilities of learned representations are heavily dependent on inductive biases which can be understood as limiting the search space of possible hypotheses a priori. Without any inductive bias, learning generalizations beyond the training data is theoretically impossible [1]. Modern machine learning researchers have adopted many task-specific inductive biases almost by default, such as convolution for spatially structured data. The field of structured representation learning has thus emerged to incorporate specific inductive biases into deep neural networks, with the hope that these structures could help machine learning models generalize better and become more robust and efficient [2]. Although there is an ongoing debate over what structures a 'good' representation should possess, there are a few generally accepted desired properties, including statistical independence and controllability [3], output symmetries [4, 5], sparse coding [6], and causality [7].

A prime example of these beneficial inductive biases is group equivariance [5]; in the case of translation, this equivariance structure results in the convolutional neural network which reduces the sensitivity of fully connected artificial neural networks to small image shifts and deformations. The adoption of the architectural motif of convolution, and thereby translation equivariance, was arguably the driving factor behind the rapid growth of the field of deep learning in the early 2010s [8]. Another closely related line of work has focused on developing representations which respect symmetries of the underlying data in their output space [4, 9]. Specifically, equivariant representations are those for which the output transforms in a known predictable way for a given input transformation. They can be seen to share many similarities with disentangled representations since an object undergoing a transformation which preserves its identity can be called a symmetry transformation [9]. This book mainly focuses on the study of learning the disentangled representations which are simultaneously approximately equivariant. In the following sections, we will first introduce disentangled representation learning, then proceed to discuss more strictly defined equivariant neural networks, and end with the intersection between these two fields—approximately equivariant and disentangled representations.

2.2 Disentangled Representation Learning

The idea of learning disentangled representations dates back to factorizing non-redundant input patterns [10] but was most recently popularized in the deep learning community by early models such as InfoGAN [11] and the β-VAE [3]. InfoGAN [11] disentangles the latent space by maximizing the mutual information between a subset of latent dimensions and observations, while β-VAE [3] factorizes the posterior $q(\mathbf{z}|\mathbf{x})$ by penalizing the total correlation between the prior and variational posterior. Subsequent work following InfoGAN mainly focused on discovering different semantically interpretable directions in the latent space [12–26]. Goetschalckx et al. [12] and Jahanian et al. [13] analyzed the steerability in the latent spaces of GANs and the impact on the properties of interests. Peebles et al. [17] and Wei et al. [19] proposed to disentangle the latent space through regularizing the Hessian and Jacobian matrices with respect to latent variables, respectively. Voynov et al. [14] first proposed an unsupervised framework for discovering disentangled latent directions in pre-trained GANs for image property manipulation, which inspired a rich line of research on discovering semantically meaningful disentangled latent directions [18, 23, 26]. More recently, some works [20, 22, 24] leverage the disentangled directions and matrix decomposition techniques to perform local editing of images.

Following β-VAE, many attempts have been made to encourage independence of the aggregated posterior through additional guidance [27–38]. β-TCVAE [31] further decompose the KL divergence term into three components and assign them with different penalization strengths. Kim et al. [30] proposed FactorVAE which imposes independence constraints by penalizing the total correlation of single latent dimensions. JointVAE et al. [28] proposed a framework which is capable of disentangling both continuous representations and discrete factors of variations in an unsupervised manner. SlowVAE [39] provided evidence that a sparse distribution characterizes the temporal transitions of natural videos and therefore proposed a sparsity prior for disentangled representation learning. With the recent powerful score-based generative models, some research efforts proposed to disentangle denoising diffusion models by crafting compact low-dimensional latent spaces [40–43]. For example,

Asyrp [40] proposed to consider the bottleneck of the denoising model as the semantic latent space due to the nice properties that accommodate semantic image manipulation.

On the other hand, many works strive to improve the understanding of disentangled representations from various perspectives. Higgens et al. [9] and Cohen et al. [5] studied the mathematical definitions of disentangled representations through group theory. Bouchacourt et al. [44] addressed the topological defects of disentanglement through distributed latent operators acting on the entire latent space. Some works even drew connections of disentangled representations to brain activities [45, 46].

Another important research branch of disentangled representationes is called sequential disentanglement [47–60] where the disentangled representation learning techniques are applied to sequence data like video and audio. In the sequential case, latent variables are typically split into single static time-invariant codes that do not change over time and multiple dynamic time-varying components that describe the distinct motions in the sequence. Due to the static and dynamic assumptions, these methods have to use two sets of latent variables for modeling different components.

2.3 Equivariant Neural Networks

A function is said to be an equivariant map if it commutes with a given transformation, i.e., $T'[f(x)] = f(T[x])$ where T and T' represent operators in different domains. Equivariance has been considered a desired inductive bias for deep neural networks as this property can preserve geometric symmetries of the input space [61–65]. Analytically equivariant networks typically enforce explicit symmetry to group transformations in neural networks [2, 5, 66–71]. Cohen et al. [5, 66, 72] successively developed group convolution layers, steerable representations, and spherical cross-correlations to build rotation-equivariant convolution neural networks for classifying images. Beyond image recognition, other important application scenarios of equivariant models include permutation-equivariant graph neural networks allowing for graph/set processing [73, 74], SE(3)-equivariant models for protein/molecule/RNA prediction [71, 75, 76], and physics-informed neural networks are equivariant under the symmetries of certain physical systems [77, 78]. We kindly refer readers to [79, 80] for a comprehensive introduction to equivariant deep learning.

Compared with disentanglement methods, equivariant networks are much more strictly defined, allowing for significantly greater control and theoretical guarantees with respect to the learned transformation [5, 71, 81–83]. However, this restriction also limits the types of transformations to which they may be applied. For example, currently only group transformations are supported, limiting real-world applicability. To avoid this caveat, some recent attempts propose to learn general but approximate equivariance from disentangled representations [25, 39, 84], which we will detail in the next section.

2.4 Approximately Equivariant and Disentangled Representations

Given the strict limitation of analytical equivariant networks, some works proposed to relax the group constraints for learning more flexible equivariant representations.

2.4.1 Prior Work: Capsule Networks, Homeomorphic VAEs

The idea of equivariant representations is that symmetry transformations define equivalence classes as the orbits of their transformations, and this structure is expected to be maintained in the deeper layers of a neural network. For instance, for images, asserting a rotated image contains the same object for all rotations, the rotation transformation then defines an orbit where the elements of that orbit can be interpreted as pose or angular orientations. When an image is processed by a neural network, we want features at different orientations to be able to be combined to form new features, but we want to ensure the relative pose information between the features is preserved for all orientations. This has the advantage that the equivalence class of rotations for the complex composite features is guaranteed to be maintained, allowing for the extraction of invariant features, a unified pose, and increased data efficiency. Such ideas are reminiscent of the capsule networks of Hinton et al. [61, 85, 86] where each capsule should represent a specific type of entity such as an object or an object part. The formal connections to equivariance have further been made in [87] by developing group capsules and the aggregation algorithm. Interestingly, by explicitly building neural networks to be equivariant, we additionally see geometric organization of activations into these equivalence classes, and further, the elements within an equivalence class are seen to exhibit higher-order non-Gaussian dependencies [88–91]. On the other hand, homeomorphic VAEs [92] induced the topological organizations to the manifold of latent variables by enforcing a homeomorphic map of SO(3) groups and showed that the manifold-valued latent variables are able to learn group actions.

2.4.2 Biological and Physical Inductive Biases for Learning Equivariant Representations

Similar to the inductive biases in machine learning, natural intelligence as implemented by biological systems also has many inductive biases by virtue of the diversity of constraints that it must simultaneously satisfy, such as metabolic efficiency and complex homeostasis. Scientific disciplines such as psychology, cognitive science, and neuroscience have all studied these biases and their observed signatures, often hypothesizing about their computational implications. Moreover, the world around us obeys certain physical laws such as the conservation of mass and momentum, as well as continuity equations which ensure transition

2.4 Approximately Equivariant and Disentangled Representations

smoothness. In attempting to build a machine learning model which is able to represent such a world, it would therefore intuitively seem beneficial to integrate similar inductive biases into the models a priori. In this book, we draw inspiration from various natural intelligent systems and physical laws to enforce beneficial inductive biases for learning approximate equivariance.

Many parts of the brain are organized topographically. Famous examples are the ocular dominance maps and the orientation maps in V1. One potential explanation for this topographic organizations is believed to be the principle of redundancy reduction, which strives to represent information as efficiently and statistically independent as possible. As discussed in Sect. 2.4.1, the prior work on Homeomorphical VAEs and capsule networks hints at the possibility of encouraging approximate equivariance from an induced topology in feature space. In Chap. 3, motivated by the connection between topographic organizations and equivariance, we will present a structured VAE where the topographic organization of observed transformations is induced, and features of these transformations are grouped into equivariant capsules.

In the neuroscience community, traveling waves have been measured at a diversity of regions and scales in the brain, however a consensus as to their computational purpose has yet to be reached. An intriguing hypothesis is that traveling waves serve to structure neural representations both in space and time, thereby acting as an inductive bias toward natural data which exhibits a similar space-time correlation structure. In Chap. 4, we propose to investigate the computational role of spatiotemporal dynamics by biasing a recurrent network to towards the production of traveling waves in its hidden state. We demonstrate experimentally that this new model indeed learns spatial structures such as topographic organizations and further provably uses this structure to encode observed transformations.

Furthermore, there is a variety of emerging work suggesting that traveling waves play the role of integrating information across time, encoding motion, and modulating information transfer. This implication can be seen similar to the latent traversal of generative models, i.e., steering latent variable for disentangled controls of output variations. Following the previous hypothesis on the role of traveling waves, Chap. 5 takes a step further to investigate if traveling waves could be a neural correlate of so called 'latent traversals' in artificial neural networks. Ultimately, we show that adding wave-like dynamics as a prior on neural dynamics indeed benefits generative models for discovering more independent steerable directions in latent space, corresponding to semantic transformations in the generated outputs.

In Chap. 6, we turn to leverage physical constraints to build approximately equivariant models. Inspired by the dynamic Optimal Transport (OT) interpretation in fluid mechanics, we propose to model the probability flow paths of observed transformations as distinct fluid motions that follow OT. This novel framework allows for a novel perspective on both *disentanglement* and *generalized equivariance*. The novel definition of disentanglement refers to the distinct set of tangent directions that follow the OT paths for modeling different factors of variation. The concept of equivariance means that the two probabilistic paths in the image space and in the latent space would eventually result in the same distribution of transformed data.

The framework we present in Chap. 6 is promising but requires supervision or weak supervision of the pure transformation sequences. Chapter 7 targets this drawback and proposes two natural-world-inspired improvements to avoid any sort of supervision. Motivated by the temporal sparsity in natural videos, we use spike-and-slab priors to model the time-varying transformation category and speeds. We also leverage the Helmholtz decomposition to factorize the latent flow fields into curl-free and divergence-free vector fields. This decomposition is principled in physics and helps us to understand the dynamics of real-world transformations.

References

1. David H Wolpert. The lack of a priori distinctions between learning algorithms. *Neural computation*, 8(7):1341–1390, 1996.
2. Daniel E Worrall, Stephan J Garbin, Daniyar Turmukhambetov, and Gabriel J Brostow. Harmonic networks: Deep translation and rotation equivariance. In *CVPR*, 2017.
3. Irina Higgins, Loic Matthey, Arka Pal, Christopher Burgess, Xavier Glorot, Matthew Botvinick, Shakir Mohamed, and Alexander Lerchner. beta-vae: Learning basic visual concepts with a constrained variational framework. *ICLR*, 2016.
4. Taco S Cohen and Max Welling. Transformation properties of learned visual representations. *ICLR*, 2015.
5. Taco Cohen and Max Welling. Group equivariant convolutional networks. In *ICML*, 2016.
6. Bruno A Olshausen and David J Field. Sparse coding with an overcomplete basis set: A strategy employed by V1? *Vision research*, 37(23):3311–3325, 1997.
7. Bernhard Schölkopf, Francesco Locatello, Stefan Bauer, Nan Rosemary Ke, Nal Kalchbrenner, Anirudh Goyal, and Yoshua Bengio. Toward causal representation learning. *Proceedings of the IEEE*, 2021.
8. Alex Krizhevsky, Ilya Sutskever, and Geoffrey E Hinton. Imagenet classification with deep convolutional neural networks. *Advances in neural information processing systems*, 25, 2012.
9. Irina Higgins, David Amos, David Pfau, Sebastien Racaniere, Loic Matthey, Danilo Rezende, and Alexander Lerchner. Towards a definition of disentangled representations. *arXiv preprint* arXiv:1812.02230, 2018.
10. Jürgen Schmidhuber. Learning factorial codes by predictability minimization. *Neural computation*, 1992.
11. Xi Chen, Yan Duan, Rein Houthooft, John Schulman, Ilya Sutskever, and Pieter Abbeel. Infogan: Interpretable representation learning by information maximizing generative adversarial nets. *NeurIPS*, 2016.
12. Lore Goetschalckx, Alex Andonian, Aude Oliva, and Phillip Isola. Ganalyze: Toward visual definitions of cognitive image properties. In *ICCV*, 2019.
13. Ali Jahanian, Lucy Chai, and Phillip Isola. On the "steerability" of generative adversarial networks. *ICLR*, 2020.
14. Andrey Voynov and Artem Babenko. Unsupervised discovery of interpretable directions in the gan latent space. In *ICML*, 2020.
15. Erik Härkönen, Aaron Hertzmann, Jaakko Lehtinen, and Sylvain Paris. Ganspace: Discovering interpretable gan controls. *NeurIPS*, 2020.
16. Xinqi Zhu, Chang Xu, and Dacheng Tao. Learning disentangled representations with latent variation predictability. In *ECCV*, 2020.

17. William Peebles, John Peebles, Jun-Yan Zhu, Alexei Efros, and Antonio Torralba. The hessian penalty: A weak prior for unsupervised disentanglement. In *ECCV*, 2020.
18. Yujun Shen and Bolei Zhou. Closed-form factorization of latent semantics in gans. In *CVPR*, 2021.
19. Yuxiang Wei, Yupeng Shi, Xiao Liu, Zhilong Ji, Yuan Gao, Zhongqin Wu, and Wangmeng Zuo. Orthogonal jacobian regularization for unsupervised disentanglement in image generation. In *ICCV*, 2021.
20. Jiapeng Zhu, Ruili Feng, Yujun Shen, Deli Zhao, Zheng-Jun Zha, Jingren Zhou, and Qifeng Chen. Low-rank subspaces in gans. *NeurIPS*, 2021.
21. Christos Tzelepis, Georgios Tzimiropoulos, and Ioannis Patras. WarpedGANSpace: Finding non-linear rbf paths in GAN latent space. In *ICCV*, 2021.
22. Jiapeng Zhu, Yujun Shen, Yinghao Xu, Deli Zhao, and Qifeng Chen. Region-based semantic factorization in gans. *ICML*, 2022.
23. Yue Song, Nicu Sebe, and Wei Wang. Orthogonal svd covariance conditioning and latent disentanglement. *IEEE T-PAMI*, 2022.
24. James Oldfield, Christos Tzelepis, Yannis Panagakis, Mihalis A Nicolaou, and Ioannis Patras. Panda: Unsupervised learning of parts and appearances in the feature maps of gans. *ICLR*, 2023.
25. Yue Song, Andy Keller, Nicu Sebe, and Max Welling. Latent traversals in generative models as potential flows. In *ICML*. PMLR, 2023.
26. Yue Song, Jichao Zhang, Nicu Sebe, and Wei Wang. Householder projector for unsupervised latent semantics discovery. In *ICCV*, 2023.
27. Nat Dilokthanakul, Pedro AM Mediano, Marta Garnelo, Matthew CH Lee, Hugh Salimbeni, Kai Arulkumaran, and Murray Shanahan. Deep unsupervised clustering with gaussian mixture variational autoencoders. *ICLR*, 2016.
28. Emilien Dupont. Learning disentangled joint continuous and discrete representations. *NeurIPS*, 2018.
29. Abhishek Kumar, Prasanna Sattigeri, and Avinash Balakrishnan. Variational inference of disentangled latent concepts from unlabeled observations. *ICLR*, 2018.
30. Hyunjik Kim and Andriy Mnih. Disentangling by factorising. In *ICML*, 2018.
31. Ricky TQ Chen, Xuechen Li, Roger B Grosse, and David K Duvenaud. Isolating sources of disentanglement in variational autoencoders. *NeurIPS*, 2018.
32. Yeonwoo Jeong and Hyun Oh Song. Learning discrete and continuous factors of data via alternating disentanglement. In *ICML*, 2019.
33. Cagatay Yildiz, Markus Heinonen, and Harri Lahdesmaki. Ode2vae: Deep generative second order odes with bayesian neural networks. *NeurIPS*, 2019.
34. Zheng Ding, Yifan Xu, Weijian Xu, Gaurav Parmar, Yang Yang, Max Welling, and Zhuowen Tu. Guided variational autoencoder for disentanglement learning. In *CVPR*, 2020.
35. Huajie Shao, Shuochao Yao, Dachun Sun, Aston Zhang, Shengzhong Liu, Dongxin Liu, Jun Wang, and Tarek Abdelzaher. Controlvae: Controllable variational autoencoder. In *ICML*. PMLR, 2020.
36. Francesco Locatello, Ben Poole, Gunnar Rätsch, Bernhard Schölkopf, Olivier Bachem, and Michael Tschannen. Weakly-supervised disentanglement without compromises. In *ICML*. PMLR, 2020.
37. Chang-Yu Tai, Ming-Yao Li, and Lun-Wei Ku. Hyperbolic disentangled representation for fine-grained aspect extraction. In *AAAI*, 2022.
38. Benjamin Estermann and Roger Wattenhofer. Dava: Disentangling adversarial variational autoencoder. *ICLR*, 2023.
39. David Klindt, Lukas Schott, Yash Sharma, Ivan Ustyuzhaninov, Wieland Brendel, Matthias Bethge, and Dylan Paiton. Towards nonlinear disentanglement in natural data with temporal sparse coding. *ICLR*, 2021.

40. Mingi Kwon, Jaeseok Jeong, and Youngjung Uh. Diffusion models already have a semantic latent space. *ICLR*, 2023.
41. Yong-Hyun Park, Mingi Kwon, Jaewoong Choi, Junghyo Jo, and Youngjung Uh. Understanding the latent space of diffusion models through the lens of riemannian geometry. *NeurIPS*, 2023.
42. Tao Yang, Yuwang Wang, Yan Lv, and Nanning Zh. Disdiff: Unsupervised disentanglement of diffusion probabilistic models. *NeurIPS*, 2023.
43. Yingheng Wang, Yair Schiff, Aaron Gokaslan, Weishen Pan, Fei Wang, Christopher De Sa, and Volodymyr Kuleshov. Infodiffusion: Representation learning using information maximizing diffusion models. In *ICML*. PMLR, 2023.
44. Diane Bouchacourt, Mark Ibrahim, and Stéphane Deny. Addressing the topological defects of disentanglement via distributed operators. *arXiv preprint* arXiv:2102.05623, 2021.
45. Irina Higgins, Le Chang, Victoria Langston, Demis Hassabis, Christopher Summerfield, Doris Tsao, and Matthew Botvinick. Unsupervised deep learning identifies semantic disentanglement in single inferotemporal face patch neurons. *Nature communications*, 2021.
46. James CR Whittington, Will Dorrell, Surya Ganguli, and Timothy Behrens. Disentanglement with biological constraints: A theory of functional cell types. In *ICLR*, 2023.
47. Wei-Ning Hsu, Yu Zhang, and James Glass. Unsupervised learning of disentangled and interpretable representations from sequential data. *NeurIPS*, 2017.
48. Remi Denton and Vighnesh Birodkar. Unsupervised learning of disentangled representations from video. *NeurIPS*, 2017.
49. Ruben Villegas, Jimei Yang, Seunghoon Hong, Xunyu Lin, and Honglak Lee. Decomposing motion and content for natural video sequence prediction. *ICLR*, 2017.
50. Yingzhen Li and Stephan Mandt. Disentangled sequential autoencoder. In *ICML*. PMLR, 2018.
51. Sergey Tulyakov, Ming-Yu Liu, Xiaodong Yang, and Jan Kautz. Mocogan: Decomposing motion and content for video generation. *CVPR*, 2018.
52. Yizhe Zhu, Martin Renqiang Min, Asim Kadav, and Hans Peter Graf. S3vae: Self-supervised sequential vae for representation disentanglement and data generation. *CVPR*, 2020.
53. Sarthak Bhagat, Shagun Uppal, Zhuyun Yin, and Nengli Lim. Disentangling multiple features in video sequences using gaussian processes in variational autoencoders. *ECCV*, 2020.
54. Masanori Yamada, Heecheol Kim, Kosuke Miyoshi, Tomoharu Iwata, and Hiroshi Yamakawa. Disentangled representations for sequence data using information bottleneck principle. In *ICML*. PMLR, 2020.
55. Junwen Bai, Weiran Wang, and Carla Gomes. Contrastively disentangled sequential variational autoencoder. *NeurIPS*, 2021.
56. Jun Han, Martin Renqiang Min, Ligong Han, Li Erran Li, and Xuan Zhang. Disentangled recurrent wasserstein autoencoder. *ICLR*, 2021.
57. Sana Tonekaboni, Chun-Liang Li, Sercan Arik, Anna Goldenberg, and Tomas Pfister. Decoupling local and global representations of time series. *AISTATS*, 2022.
58. Ilan Naiman and Omri Azencot. An operator theoretic approach for analyzing sequence neural networks. *AAAI*, 2023.
59. Nimrod Berman, Ilan Naiman, and Omri Azencot. Multifactor sequential disentanglement via structured koopman autoencoders. *ICLR*, 2023.
60. Berman Nimrod, Ilan Naiman, Idan Arbiv, Gal Fadlon, and Omri Azencot. Sequential disentanglement by extracting static information from a single sequence element. In *ICML*. PMLR, 2024.
61. Geoffrey E Hinton, Alex Krizhevsky, and Sida D Wang. Transforming auto-encoders. In *ICANN*. Springer, 2011.
62. Uwe Schmidt and Stefan Roth. Learning rotation-aware features: From invariant priors to equivariant descriptors. In *CVPR*, 2012.

References

63. Chen-Yu Lee, Saining Xie, Patrick Gallagher, Zhengyou Zhang, and Zhuowen Tu. Deeply-supervised nets. In *AISTATS*. PMLR, 2015.
64. Karel Lenc and Andrea Vedaldi. Understanding image representations by measuring their equivariance and equivalence. In *CVPR*, 2015.
65. Pulkit Agrawal, Joao Carreira, and Jitendra Malik. Learning to see by moving. In *ICCV*, 2015.
66. Taco S Cohen and Max Welling. Steerable cnns. *ICLR*, 2017.
67. Siamak Ravanbakhsh, Jeff Schneider, and Barnabas Poczos. Equivariance through parameter-sharing. In *ICML*. PMLR, 2017.
68. Daniel Worrall and Max Welling. Deep scale-spaces: Equivariance over scale. *NeurIPS*, 2019.
69. Elise Van der Pol, Daniel Worrall, Herke van Hoof, Frans Oliehoek, and Max Welling. Mdp homomorphic networks: Group symmetries in reinforcement learning. *NeurIPS*, 2020.
70. Marc Finzi, Samuel Stanton, Pavel Izmailov, and Andrew Gordon Wilson. Generalizing convolutional neural networks for equivariance to lie groups on arbitrary continuous data. In *ICML*. PMLR, 2020.
71. Emiel Hoogeboom, Victor Garcia Satorras, Clément Vignac, and Max Welling. Equivariant diffusion for molecule generation in 3d. In *ICML*. PMLR, 2022.
72. Taco S Cohen, Mario Geiger, Jonas Köhler, and Max Welling. Spherical cnns. *ICLR*, 2018.
73. Manzil Zaheer, Satwik Kottur, Siamak Ravanbakhsh, Barnabas Poczos, Ruslan Salakhutdinov, and Alexander Smola. Deep sets. *Arxiv*, 2018.
74. Haggai Maron, Heli Ben-Hamu, Nadav Shamir, and Yaron Lipman. Invariant and equivariant graph networks. 2019.
75. John Jumper, Richard Evans, Alexander Pritzel, Tim Green, Michael Figurnov, Olaf Ronneberger, Kathryn Tunyasuvunakool, Russ Bates, Augustin Žídek, Anna Potapenko, et al. Highly accurate protein structure prediction with alphafold. *nature*, 2021.
76. Raphael JL Townshend, Stephan Eismann, Andrew M Watkins, Ramya Rangan, Masha Karelina, Rhiju Das, and Ron O Dror. Geometric deep learning of rna structure. *Science*, 2021.
77. George Em Karniadakis, Ioannis G Kevrekidis, Lu Lu, Paris Perdikaris, Sifan Wang, and Liu Yang. Physics-informed machine learning. *Nature Reviews Physics*, 2021.
78. Tara Akhound-Sadegh, Laurence Perreault-Levasseur, Johannes Brandstetter, Max Welling, and Siamak Ravanbakhsh. Lie point symmetry and physics-informed networks. *NeurIPS*, 2023.
79. Michael M. Bronstein, Joan Bruna, Taco S Cohen, and Petar Veličković. Geometric deep learning: Grids, groups, graphs, geodesics, and gauges. *Arxiv*, 2021.
80. Maurice Weiler, Patrick Forré, Erik Verlinde, and Max Welling. *Equivariant and Coordinate Independent Convolutional Networks*. 2023.
81. Jonas Köhler, Leon Klein, and Frank Noé. Equivariant flows: exact likelihood generative learning for symmetric densities. In *ICML*. PMLR, 2020.
82. Victor Garcia Satorras, Emiel Hoogeboom, Fabian B Fuchs, Ingmar Posner, and Max Welling. E (n) equivariant normalizing flows. *NeurIPS*, 2021.
83. Neel Dey, Antong Chen, and Soheil Ghafurian. Group equivariant generative adversarial networks. *ICLR*, 2021.
84. T Anderson Keller and Max Welling. Topographic vaes learn equivariant capsules. *NeurIPS*, 2021.
85. Geoffrey E Hinton, Sara Sabour, and Nicholas Frosst. Matrix capsules with em routing. In *International conference on learning representations*, 2018.
86. Sara Sabour, Nicholas Frosst, and Geoffrey E Hinton. Dynamic routing between capsules. *Advances in neural information processing systems*, 30, 2017.
87. Jan Eric Lenssen, Matthias Fey, and Pascal Libuschewski. Group equivariant capsule networks. In *NeurIPS*, pages 8858–8867, 2018.
88. Siwei Lyu and Eero P Simoncelli. Nonlinear image representation using divisive normalization. In *2008 IEEE Conference on Computer Vision and Pattern Recognition*, pages 1–8. IEEE, 2008.

89. Siwei Lyu and Eero P Simoncelli. Modeling multiscale subbands of photographic images with fields of gaussian scale mixtures. *IEEE Transactions on pattern analysis and machine intelligence*, 31(4):693–706, 2008.
90. Martin J Wainwright and Eero Simoncelli. Scale mixtures of gaussians and the statistics of natural images. *Advances in neural information processing systems*, 12, 1999.
91. Martin J Wainwright, Eero P Simoncelli, and Alan S Willsky. Random cascades on wavelet trees and their use in analyzing and modeling natural images. *Applied and Computational Harmonic Analysis*, 11(1):89–123, 2001.
92. Luca Falorsi, Pim De Haan, Tim R Davidson, Nicola De Cao, Maurice Weiler, Patrick Forré, and Taco S Cohen. Explorations in homeomorphic variational auto-encoding. *arXiv preprint* arXiv:1807.04689, 2018.

Part II
Naturally Inspired Topographically Structured Representation Learning

Topographic Variational Autoencoders

3.1 Introduction

Many parts of the brain are organized topographically. Famous examples are the ocular dominance maps and the orientation maps in V1. What is the advantage of such organization and what can we learn from it to develop better inductive biases for deep neural network architectures?

One potential explanation for the emergence of topographic organization is provided by the principle of redundancy reduction [1]. In the language of Information Theory, redundancy wastes channel capacity, and thus to represent information as efficiently as possible, the brain may strive to transform the input to a neural code where the activations are statistically maximally independent. In the machine learning literature, this idea resulted in Independent Component Analysis (ICA) which linearly transforms the input to a new basis where the activities are independent and sparse [2–5]. It was soon realized that there are remaining higher order dependencies (such as correlation between absolute values) that can not be transformed away by a linear transformation. For example, along edges of an image, linear-ICA components (e.g. gabor filters) still activate in clusters even though the sign of their activity is unpredictable [6, 7]. This led to new algorithms that explicitly model these remaining dependencies through a topographic organization of feature activations [8–11]. Such topographic features were reminiscent of pinwheel structures observed in V1, encouraging multiple comparisons with topographic organization in the biological visual system [12–14] (Fig. 3.1).

A second, almost independent body of literature developed the idea of "equivariance" of neural network feature maps under symmetry transformations. Interestingly, by explicitly building neural networks to be equivariant, there are guaranteed to be higher-order non-Gaussian dependencies between the tied-weights of the model, analagous to those found in the aforementioned topographic models [7, 15–17]. The insight of this connection between

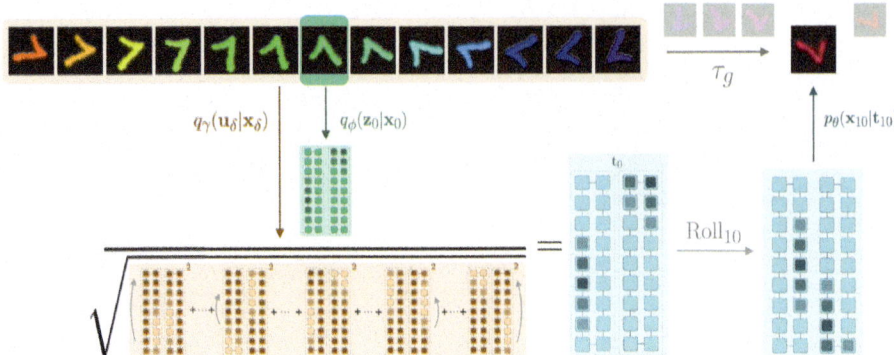

Fig. 3.1 Overview of the Topographic VAE with shifting temporal coherence. The combined color/rotation transformation in input space τ_g becomes encoded as a Roll within the capsule dimension. The model is thus able decode unseen sequence elements by encoding a partial sequence and Rolling activations within the capsules. We see this resembles a commutative diagram

topographic organization and equivariance hints at a possibility to encourage approximate equivariance from an induced topology in feature space.

To build a model, we need to ask what mechanisms could induce topographic organization of *observed transformations* specifically? We have argued that removing dependencies between latent variables is a possible mechanism; however, to obtain the more structured organisation of equivariant capsule representations, the usual approach is to hard-code this structure into the network, or to encourage it through regularization terms [18, 19]. To achieve this same structure *unsupervised*, we propose to incorporate another key inductive bias: "temporal coherence" [20–23]. The principle of temporal coherence, or "slowness", asserts than when processing correlated sequences, we wish for our representations to change smoothly and slowly over space and time. Thinking of time sequences as symmetry transformations on the input, we desire features undergoing such transformations to be grouped into equivariant capsules. We therefore suggest that encouraging slow feature transformations to take place *within a capsule* could induce such grouping from sequences alone.

In the following sections we will explain the details of our Topographic Variational Autoencoder which lies at the intersection of topographic organization, equivariance, and temporal coherence, thereby learning approximately equivariant capsules from sequence data completely unsupervised.

3.2 The Generative Model

The generative model proposed in this chapter is based on the Topographic Product of Student's-t (TPoT) model as developed in [10, 11]. In the following, we will show how a TPoT random variable can be constructed from a set of independent univariate standard

3.2 The Generative Model

normal random variables, enabling efficient training through variational inference. Subsequently, we will construct a new model where topographic neighborhoods are extended over time, introducing temporal coherence and encouraging the unsupervised learning of approximately equivariant subspaces we call 'capsules'.

3.2.1 Topographic Generative Models

Inspired by Topographic ICA, the class of Topographic Generative models can be understood as generative models where the joint distribution over latent variables does not factorize into entirely independent factors, as is commonly done in ICA or VAEs, but instead has a more complex 'local' correlation structure. The locality is defined by arranging the latent variables into an n-dimensional lattice or grid, and organizing variables such that those which are closer together on this grid have greater correlation of activities than those which are further apart. In the related literature, activations which are nearby in this grid are defined to have higher-order correlation, e.g. correlations of squared activations (aka 'energy'), asserting that all first order correlations are removed by the initial ICA de-mixing matrix.

Such generative models can be seen as hierarchical generative models where there exist higher level independent 'variance generating' variables $\mathbf{V}$ which are combined locally to generate the variances $\sigma = \phi(\mathbf{WV})$ of the lower level topographic variables $\mathbf{T} \sim \mathcal{N}(\mathbf{0}, \sigma^2 \mathbf{I})$, for an appropriate non-linearity ϕ. The variables $\mathbf{T}$ are thus independent conditioned on σ. Other related models which can be described under this umbrella include *Independent Subspace Analysis* (ISA) [24] where all variables within a predefined subspace (or 'capsule') share a common variance, and '*temporally coherent*' models [21] where the energy of a given variable between time steps is correlated by extending the topographic neighborhoods over the time dimension [25]. The topographic latent variable $\mathbf{T}$ can additionally be described as an instance of a Gaussian scale mixture (GSM). GSMs have previously been used to model the observed non-Gaussian dependencies between coefficients of steerable wavelet pyramids (interestingly also equivariant to translation & rotation) [6, 7, 17].

3.2.2 The Product of Student's-t Model

To construct the Topographic Variational Autoencoder specifically, we assume that that our observed data is generated by a latent variable model where the joint distribution over observed and latent variables $\mathbf{x}$ and $\mathbf{t}$ factorizes into the product of the conditional and the prior. The prior distribution $p_\mathbf{T}(\mathbf{t})$ is assumed to be a Topographic Product of Student's-t (TPoT) distribution, and we parameterize the conditional distribution with a flexible function approximator:

$$p_{\mathbf{X},\mathbf{T}}(\mathbf{x}, \mathbf{t}) = p_{\mathbf{X}|\mathbf{T}}(\mathbf{x}|\mathbf{t}) p_\mathbf{T}(\mathbf{t}) \quad p_{\mathbf{X}|\mathbf{T}}(\mathbf{x}|\mathbf{t}) = p_\theta(\mathbf{x}|g_\theta(\mathbf{t})) \quad p_\mathbf{T}(\mathbf{t}) = \text{TPoT}(\mathbf{t}; \nu) \quad (3.1)$$

The goal of training is thus to learn the parameters θ such that the marginal distribution of the model $p_\theta(\mathbf{x})$ matches that of the observed data. Unfortunately, the marginal likelihood is generally intractable except for all but the simplest choices of g_θ and $p_\mathbf{T}$ [9]. Prior work has therefore resorted to techniques such as contrastive divergence with Gibbs sampling [11] to train TPoT models as energy based models. In the following section, we instead demonstrate how TPoT variables can be constructed as a deterministic function of Gaussian random variables, enabling the use of variational inference and efficient maximization of the likelihood through the evidence lower bound (ELBO).

3.2.3 Constructing the Product of Student's-t Distribution

First, note a univariate Student's-t random variable T with ν degrees of freedom can be defined as:

$$T = \frac{Z}{\sqrt{\frac{1}{\nu}\sum_i^\nu U_i^2}} \quad \text{with} \quad Z, U_i \sim \mathcal{N}(0,1) \ \forall i \tag{3.2}$$

where Z and $\{U_i\}_{i=1}^\nu$ are independent standard normal random variables. If $\mathbf{T}$ is a multidimensional Student's-t random variable, composed of independent Z_i and U_i, then $\mathbf{T} \sim \text{PoT}(\nu)$, i.e.:

$$\mathbf{T} = \left[\frac{Z_1}{\sqrt{\frac{1}{\nu}\sum_{i=1}^\nu U_i^2}}, \frac{Z_2}{\sqrt{\frac{1}{\nu}\sum_{i=\nu+1}^{2\cdot\nu} U_i^2}}, \ldots \frac{Z_n}{\sqrt{\frac{1}{\nu}\sum_{i=(n-1)\cdot\nu+1}^{n\cdot\nu} U_i^2}} \right] \sim \text{PoT}(\nu) \tag{3.3}$$

Note that the Student's-t variable T is large when most of the $\{U_i\}_i$ in its set are small. We can therefore think of the $\{U_i\}_i$ as constraint violations rather then pattern matches: if the input matches all constraints $U_i \approx 0$, the corresponding T variables will activate (see [26] for further discussion).

3.2.4 Introducing Topography

To make the PoT distribution topographic, we strive to correlate the scales of T_j which are 'nearby' in our topographic layout. One way to accomplish this is by *sharing* some U_i-variables between neighboring T_j's. Formally, we define overlapping neighborhoods $N(j)$ for each variable T_j and write:

$$\mathbf{T} = \left[\frac{Z_1}{\sqrt{\frac{1}{\nu}\sum_{i\in N(1)} U_i^2}}, \frac{Z_2}{\sqrt{\frac{1}{\nu}\sum_{i\in N(2)} U_i^2}}, \ldots \frac{Z_n}{\sqrt{\frac{1}{\nu}\sum_{i\in N(n)} U_i^2}} \right] \sim \text{TPoT}(°) \tag{3.4}$$

3.2 The Generative Model

With some abuse of notation, if we define $\mathbf{W}$ to be the adjacency matrix which defines our neighborhood structure, $\mathbf{U}$ and $\mathbf{Z}$ to be the vectors of random variables U_i and Z_j, we can write the above succinctly as:

$$\mathbf{T} = \left[\frac{Z_1}{\sqrt{\frac{1}{\nu} W_1 \mathbf{U}^2}}, \frac{Z_2}{\sqrt{\frac{1}{\nu} W_2 \mathbf{U}^2}}, \ldots \frac{Z_n}{\sqrt{\frac{1}{\nu} W_n \mathbf{U}^2}} \right] = \frac{\mathbf{Z}}{\sqrt{\frac{1}{\nu} \mathbf{W} \mathbf{U}^2}} \sim \text{TPoT}(^\circ) \quad (3.5)$$

Due to non-linearities such as ReLUs which may alter input distributions, it is beneficial to allow the Z variables to model the mean and scale. We found this can be achieved with the following parameterization: $\mathbf{T} = \frac{\mathbf{Z} - \mu}{\sigma \sqrt{1/\nu \mathbf{W} \mathbf{U}^2}}$. In practice, we found that $\sigma = \sqrt{\nu}$ often works well, finally yielding:

$$\mathbf{T} = \frac{\mathbf{Z} - \mu}{\sqrt{\mathbf{W} \mathbf{U}^2}} \quad (3.6)$$

Given this construction, we observe that the TPoT generative model can instead be viewed as a latent variable model where all random variables are Gaussian and the construction of $\mathbf{T}$ in Eq. 3.6 is the first layer of the generative 'decoder': $g_\theta(\mathbf{t}) = g_\theta(\mathbf{u}, \mathbf{z})$. In Sect. 3.3 we then leverage this interpretation to show how an approximate posterior for the latent variables $\mathbf{Z}$ and $\mathbf{U}$ can be trained through variational inference.

3.2.5 Capsules as Disjoint Topologies

One setting of neighborhood structure $\mathbf{W}$ which is of particular interest is when there exist multiple sets of disjoint neighborhoods. Statistically, the variables of two disjoint topologies are completely independent. An example of a capsule neighborhood structure is shown in Fig. 3.2. The idea of independant subspaces has previously been shown to learn invariant feature subspaces in the linear setting and is present in early work on Independent Subspace Analysis [24] and Adaptive Subspace Self Organizing Maps (ASSOM) [27]. It is also very reminiscent of the transformed sets of features present in a group equivariant convolutional neural network. In the next section, we will show how temporal coherence can be leveraged to induce the encoding of observed transformations into the internal dimensions of such capsules thereby yielding unsupervised approximately equivariant capsules.

3.2.6 Temporal Coherence and Learned Equivariance

We now describe how the induced topographic organization can be leveraged to learn a basis of approximately equivariant capsules for observed transformation sequences. The resulting representation is composed of a large set of 'capsules' where the dimensions inside the capsule are topographically structured, but between the capsules there is independence.

Fig. 3.2 An example of a neighborhood structure which induces disjoint topologies (A.K.A. capsules). Lines between variables T_i indicate that sharing of U_i, and thus correlation

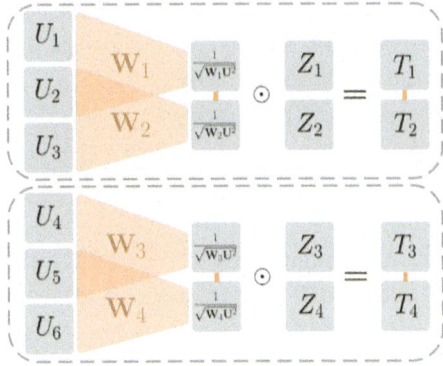

To benefit from sequences of input, we encourage topographic structure over time between sequentially permuted activations within a capsule, a property we refer to as *shifting temporal coherence*.

3.2.6.1 Temporal Coherence

Temporal Coherence can be measured as the correlation of squared activation between time steps. One way we can achieve this in our model is by having T_j share U_i between time steps. Formally, the generative model is identical to Eq. 3.1, factorizing over timesteps denoted by subscript l, i.e. $p_{\mathbf{X}_l, \mathbf{T}_l}(\mathbf{x}_l, \mathbf{t}_l) = p_{\mathbf{X}_l | \mathbf{T}_l}(\mathbf{x}_l | \mathbf{t}_l) p_{\mathbf{T}_l}(\mathbf{t}_l)$. However, $\mathbf{T}_l$ is now a function of a sequence $\{\mathbf{U}_{l+\delta}\}_{\delta=-L}^{L}$:

$$\mathbf{T}_l = \frac{\mathbf{Z}_l - \mu}{\sqrt{\mathbf{W}\left[\mathbf{U}_{l+L}^2; \cdots ; \mathbf{U}_{l-L}^2\right]}} \qquad (3.7)$$

where $\left[\mathbf{U}_{l+L}^2; \cdots ; \mathbf{U}_{l-L}^2\right]$ denotes vertical concatenation of the column vectors $\mathbf{U}_l$, and $2L$ can be seen as the window size. We see that the choice of $\mathbf{W}$ now defines correlation structure over time. In prior work on temporal coherence (denoted 'Bubbles' [25]), the grouping over time is such that a given variable $T_{l,i}$ has correlated energy with *the same spatial location* (i) at a previous time step ($l-1$) (i.e. $\text{cov}(T_{l,i}^2, T_{l-1,i}^2) > 0$). This can be implemented as:

$$\mathbf{W}\left[\mathbf{U}_{l+L}^2; \cdots ; \mathbf{U}_{l-L}^2\right] = \sum_{\delta=-L}^{L} \mathbf{W}_\delta \mathbf{U}_{l+\delta}^2 \qquad (3.8)$$

where $\mathbf{W}_\delta$ defines the topography for a single timestep, and is typically the same for all timesteps.

3.2.6.2 Learned Equivariance with Shifting Temporal Coherence

In our model, instead of requiring a single location to have correlated energies over a sequence, we would like variables at sequentially permuted locations *within a capsule* to have correlated energy between timesteps ($\text{cov}(T^2_{l,i}, T^2_{l-1,i-1}) > 0$). Similarly, this can be implemented as:

$$\mathbf{W}[\mathbf{U}^2_{l+L}; \cdots ; \mathbf{U}^2_{l-L}] = \sum_{\delta=-L}^{L} \mathbf{W}_\delta \text{Roll}_\delta(\mathbf{U}^2_{l+\delta}) \quad (3.9)$$

where $\text{Roll}_\delta(\mathbf{U}^2_{l+\delta})$ denotes a cyclic permutation of δ steps along the capsule dimension. The exact implementation of Roll can be found in Sect. 3.7.9. As we will show in Sect. 3.4.3, TVAE models with such a topographic structure learn to encode observed sequence transformations as Rolls within the capsule dimension, analogous to a group equivariant neural network where τ_ρ and Roll_1 can be seen as the action of the transformation ρ on the input and output spaces respectively.

3.3 Topographic VAE

To train the parameters of the generative model θ, we use the above formulation to parameterize an approximate posterior for **t** in terms of a deterministic transformation of approximate posteriors over simpler Gaussian latent variables **u** and **z**. Explicitly:

$$q_\phi(\mathbf{z}_l|\mathbf{x}_l) = \mathcal{N}(\mathbf{z}_l; \mu_\phi(\mathbf{x}_l), \sigma_\phi(\mathbf{x}_l)\mathbf{I}) \quad p_\theta(\mathbf{x}_l|g_\theta(\mathbf{t}_l)) = p_\theta(\mathbf{x}_l|g_\theta(\mathbf{z}_l, \{\mathbf{u}_l\})) \quad (3.10)$$

$$q_\gamma(\mathbf{u}_l|\mathbf{x}_l) = \mathcal{N}(\mathbf{u}_l; \mu_\gamma(\mathbf{x}_l), \sigma_\gamma(\mathbf{x}_l)\mathbf{I}) \quad \mathbf{t}_l = \frac{\mathbf{z}_l - \mu}{\sqrt{\mathbf{W}[\mathbf{u}^2_{l+L}; \cdots ; \mathbf{u}^2_{l-L}]}} \quad (3.11)$$

We denote this model the Topographic VAE (TVAE) and optimize the parameters θ, ϕ, γ (and μ) through the ELBO, summed over the sequence length S:

$$\sum_{l=1}^{S} \mathbb{E}_{Q_{\phi,\gamma}(\mathbf{z}_l, \mathbf{u}_l|\{\mathbf{x}_l\})}\left([\log p_\theta(\mathbf{x}_l|g_\theta(\mathbf{t}_l))] - D_{KL}[q_\phi(\mathbf{z}_l|\mathbf{x}_l)||p_\mathbf{Z}(\mathbf{z}_l)]\right.$$

$$\left. - D_{KL}[q_\gamma(\mathbf{u}_l|\mathbf{x}_l)||p_\mathbf{U}(\mathbf{u}_l)]\right) \quad (3.12)$$

where $Q_{\phi,\gamma}(\mathbf{z}_l, \mathbf{u}_l|\{\mathbf{x}_l\}) = q_\phi(\mathbf{z}_l|\mathbf{x}_l)\prod_{\delta=-L}^{L} q_\gamma(\mathbf{u}_{l+\delta}|\mathbf{x}_{l+\delta})$, and $\{\cdot\}$ denotes a set over time.

3.4 Experiments

In the following experiments, we demonstrate the viability of the Topographic VAE as a novel method for training deep topographic generative models. Additionally, we quantitatively verify that shifting temporal coherence yields approximately equivariant capsules by computing an 'equivariance loss' and a correlation metric inspired by the disentanglement literature. We show that equivariant capsule models yield higher likelihood than baselines on test sequences, and qualitatively support these results with visualizations of sequences reconstructed purely from Rolled capsule activations.

3.4.1 Evaluation Methods

As depicted in +, we make use of *capsule traversals* to qualitatively visualize the transformations learned by our network. Simply, these are constructed by encoding a partial sequence into a t_0 variable, and decoding sequentially Rolled copies of this variable. Explicitly, in the top row we show the data sequence $\{x_l\}_l$, and in the bottom row we show the decoded sequence: $\{g_\theta(\text{Roll}_l(t_0))\}_l$.

To measure equivariance quantitatively, we measure an *equivariance error* similar to [19]. The equivariance error can be seen as the difference between traversing the two distinct paths of the commutative diagram, and provides some measure of how precisely the function and the transform commute. Formally, for a sequence of length S, and $\hat{t} = t/||t||_2$, the error is defined as:

$$\mathcal{E}_{eq}(\{t_l\}_{l=1}^S) = \sum_{l=1}^{S-1}\sum_{\delta=1}^{S-l} \left\| \text{Roll}_\delta(\hat{t}_l) - \hat{t}_{l+\delta} \right\|_1 \qquad (3.13)$$

Additionally, inspired by existing disentanglement metrics, we measure the degree to which observed transformations in capsule space are correlated with input transformations by introducing a new metric we call CapCorr_y. Simply, this metric computes the correlation between the amount of observed Roll of a capsule's activation at two timesteps l and $l + \delta$, and the shift of the ground truth generative factors y_l in that same time. Formally, for a correlation coefficient Corr:

$$\text{CapCorr}(t_l, t_{l+\delta}, y_l, y_{l+\delta}) = \text{Corr}\left(\text{argmax}\left[t_l \star t_{l+\delta}\right], |y_l - y_{l+\delta}|\right) \qquad (3.14)$$

where $\star$ is discrete periodic cross-correlation across the capsule dimension, and the correlation coefficient is computed across the entire dataset. We see the argmax of the cross-correlation is an estimate of the degree to which a capsule activation has shifted from time l to $l + \delta$. To extend this to multiple capsules, we can replace the argmax function with the mode of the argmax computed for all capsules. We provide additional details and extensions of this metric in Sect. 3.7.8. For measuring capsule-metrics on baseline models which do not naturally have capsules, we simply arbitrarily divide the latent space into a fixed set of corresponding capsules and capsule dimensions, and provide such results as equivalent to 'random baselines' for these metrics.

3.4.2 Topographic VAE Without Temporal Coherence

To validate the TVAE is capable of learning topographically organized representations with deep neural networks, we first perform experiments on a Topographic VAE without Temporal Coherence. The model is constructed as in Eqs. 3.10 and 3.11 with $L = 0$, and is trained to maximize Eq. 3.12. We fix $\mathbf{W}$ such that globally the latent variables are arranged in a grid on a 2-dimensional torus (a single capsule), and locally $\mathbf{W}$ sums over 5×5 2D groups of variables. In this setting, $\mathbf{W}$ can be easily implemented as 2D convolution with a 5×5 kernel of 1's, stride 1, and cyclic padding. We see that training the model with 3-layer MLP's for the encoders and decoder indeed yields a 2D topographic organization of higher level features. In Fig. 3.3, we show the maximum activating image for each final layer neuron of the capsule, plotted as a flattened torus. We see that the neurons become arranged according to class, orientation, width, and other learned features.

3.4.3 Learning Equivariant Capsules

In the remaining experiments, we provide evidence that the Topographic VAE can be leveraged to learn equivariant capsules by incorporating shifting temporal coherence into a 1D baseline topographic model. We compare against two baselines: standard normal VAEs and models that have non-shifting 'stationary' temporal coherence as defined in Eq. 3.8 (denoted 'BubbleVAE' [25]).

In all experiments we use a 3-layer MLP with ReLU activations for both encoders and the decoder. We arrange the latent space into 15 circular capsules each of 15-dimensions for dSprites [28], and 18 circular capsules each of 18-dimensions for MNIST [29]. Example sequences $\{\mathbf{x}_l\}_{l=1}^{S}$ are formed by taking a random initial example, and sequentially transforming it according to one of the available transformations: (X-Pos, Y-Pos, Orientation, Scale) for dSprites, and (Color, Scale, Orientation) for MNIST. All transformation sequences are cyclic such that when the maximum transformation parameter is reached, the subsequent value returns to the minimum. We denote the length of a full transformation sequence by S, and the time-extent of the induced temporal coherence (i.e. the length of the input sequence) by $2L$. For simplicity, both datasets are constructed such that the sequence length S equals the capsule dimension (for dSprites this involves taking a subset of the full dataset and looping the scale 3-times for a scale-sequence). Exact details are in Sects. 3.7.6 and 3.7.7.

In Fig. 3.4, we show the capsule traversals for TVAE models with $L \approx \frac{1}{3}S$. We see that despite the $\mathbf{t}_0$ variable encoding only $\frac{2}{3}$ of the sequence, the remainder of the transformation sequence can be decoded nearly perfectly by permuting the activation through the full capsule – implying the model has learned to be approximately equivariant to full sequences while only observing partial sequences per training point. Furthermore, we see that the model is able to successfully learn all transformations simultaneously for the respective datasets.

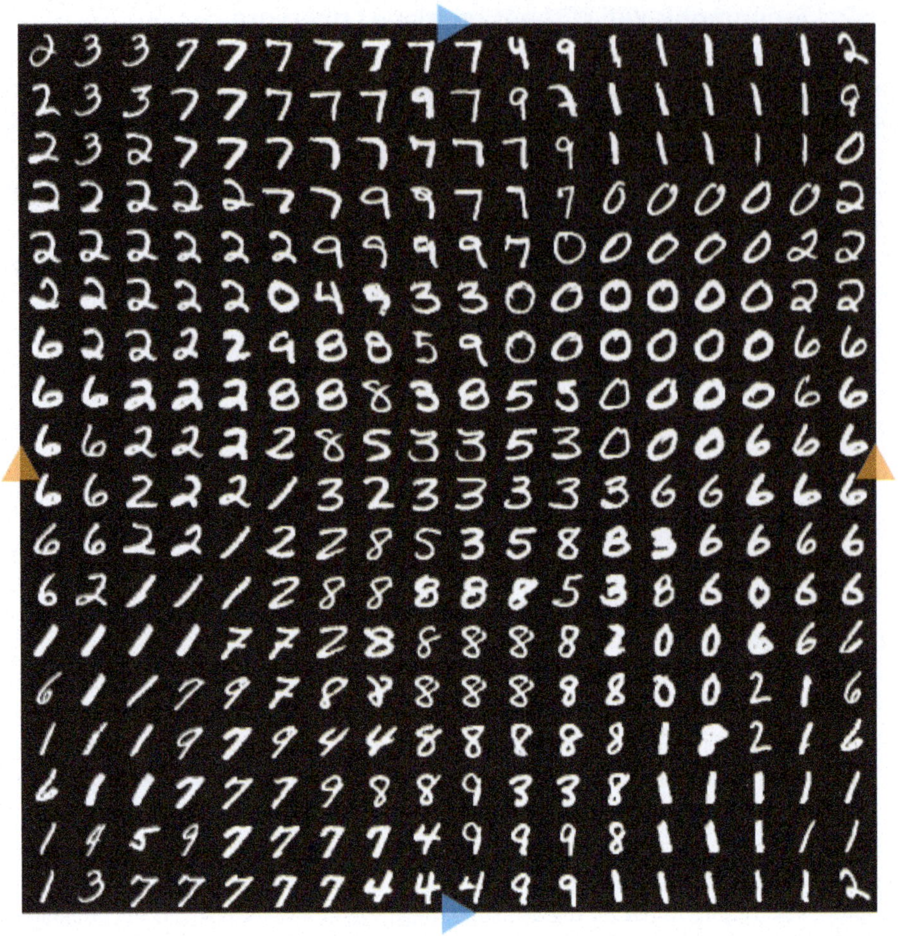

Fig. 3.3 Maximum activating images for a topographic VAE trained with a 2D torus topography on MNIST

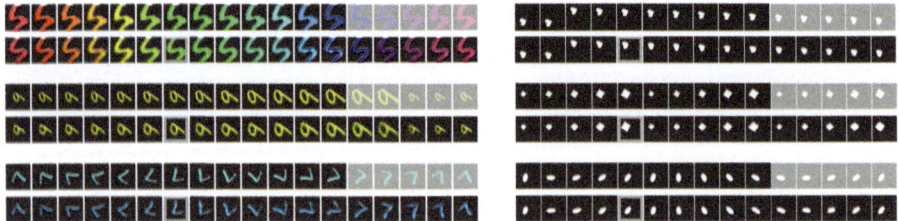

Fig. 3.4 Capsule Traversals for TVAE models on dSprites and MNIST. The top rows show the encoded sequences (with greyed-out images held-out), and the bottom rows show the images generated by decoding sequentially Rolled copies of the initial activation t_0 (indicated by a grey border)

3.4 Experiments

Table 3.1 Log Likelihood and Equivariance Error on MNIST for different settings of temporal coherence length L relative to sequence length S. Mean ± std. over 3 random initalizations

Model	TVAE	TVAE	TVAE	BubbleVAE	VAE
L	$L = \frac{1}{2}S$	$L = \frac{5}{36}S$	$L = 0$	$L = \frac{5}{36}S$	$L = 0$
$\log p(\mathbf{x}) \uparrow$	$\mathbf{-186.8 \pm 0.1}$	-186.0 ± 0.7	-218.5 ± 0.9	-191.4 ± 0.5	-189.0 ± 0.8
$\mathcal{E}_{eq} \downarrow$	$\mathbf{574 \pm 2}$	3247 ± 3	3217 ± 105	3370 ± 12	13274 ± 1

Table 3.2 Equivariance error ($\mathcal{E}_{eq} \downarrow$) and correlation of observed capsule roll with ground truth factor shift (CapCorr $\uparrow$) for the dSprites dataset. Mean ± standard deviation over 3 random initalizations

Model	TVAE	TVAE	TVAE	TVAE	BubbleVAE	VAE
L	$L = \frac{1}{2}S$	$L = \frac{1}{3}S$	$L = \frac{1}{6}S$	$L = 0$	$L = \frac{1}{3}S$	$L = 0$
CapCorr$_X \uparrow$	$\mathbf{1.0 \pm 0}$	$\mathbf{1.0 \pm 0}$	0.67 ± 0.02	0.17 ± 0.03	0.13 ± 0.01	0.18 ± 0.01
CapCorr$_Y \uparrow$	$\mathbf{1.0 \pm 0}$	$\mathbf{1.0 \pm 0}$	0.66 ± 0.02	0.21 ± 0.02	0.12 ± 0.01	0.16 ± 0.01
CapCorr$_O \uparrow$	$\mathbf{1.0 \pm 0}$	$\mathbf{1.0 \pm 0}$	0.52 ± 0.01	0.09 ± 0.01	0.10 ± 0.01	0.11 ± 0.00
CapCorr$_S \uparrow$	$\mathbf{1.0 \pm 0}$	$\mathbf{1.0 \pm 0}$	0.42 ± 0.01	0.51 ± 0.01	0.50 ± 0.00	0.52 ± 0.00
$\mathcal{E}_{eq} \downarrow$	$\mathbf{344 \pm 5}$	1034 ± 6	2549 ± 38	2971 ± 9	1951 ± 34	6934 ± 0

Capsule traversals for the non-equivariant baselines, as well as TVAEs with smaller values of L (which only learn approximate equivariance to partial sequences) are shown in Sect. 3.10. We note that the capsule traversal plotted in Fig. 3.1 demonstrates a transformation where color and rotation change simultaneously, differing from how the models in this section are trained. However, as we describe in more detail in Sect. 3.8.3, we observe that TVAEs trained with individual transformations in isolation (as in this section) are able to generalize, generating sequences of combined transformations when presented with such partial input sequences at test time. We believe this generalization capability to be promising for data efficiency, but leave further exploration to future work. Additional capsule traversals with such unseen combined transformations are shown in Sect. 3.8.3 and further complex learned transformations (such as perspective transforms) are shown at the end of Sect. 3.10.

For a more quantitative evaluation, in Table 3.1 we measure the equivariance error and log-likelihood (reported in nats) of the test data under our trained MNIST models as estimated by importance sampling with 10 samples. We observe that models which incorporate temporal coherence (BubbleVAE and TVAE with $L > 0$) achieve low equivariance error, while the TVAE models with shifting temporal coherence achieve the highest likelihood and the lowest equivariance error simultaneously.

To further understand how capsules transform for observed input transformations, in Table 3.2 we measure $\mathcal{E}_{eq}$ and the CapCorr metric on the dSprites dataset for the four proposed transformations. We see that the TVAE with $L \geq \frac{1}{3}S$ achieves perfect correlation—implying the learned representation indeed permutes cyclically within capsules for observed transformation sequences. Further, this correlation gradually decreases as L decreases, eventually reaching the same level as the baselines. We also see that, on both datasets, the equivariance losses for the TVAE with $L = 0$ and the BubbleVAE are significantly lower than the baseline VAE, while conversely, the CapCorr metric is not significantly better. We believe this to be due to the fundamental difference between the metrics: $\mathcal{E}_{eq}$ measures continuous L1 similarity which is still low when a representation is locally smooth (even if the change of the representation does not follow the observed transformation), whereas CapCorr more strictly measures the correspondence between the transformation of the input and the transformation of the representation. In other words, $\mathcal{E}_{eq}$ may be misleadingly low for invariant capsule representations (as with the BubbleVAE), whereas CapCorr strictly measures equivariance.

3.5 Future Work and Limitations

The model presented in this work has a number of limitations in its existing form which we believe to be interesting directions for future research. Foremost, the model is challenging to compare directly with existing disentanglement and equivariance literature since it requires an input sequence which determines the transformations reachable through the capsule roll. Related to this, we note the temporal coherence proposed in our model is not 'causal' (i.e. $\mathbf{t}_0$ depends on future $\mathbf{x}_l$). We believe these limitations could be at least partially alleviated with minor extensions detailed in Sect. 3.9.

We additionally note that some model developers may find a priori definition of topographic structure burdensome. While true, we know that the construction of appropriate priors is always a challenging task in latent variable models, and we observe that our proposed TVAE achieves strong performance even with improper specification. Furthermore, in future work, we believe adding learned flexibility to the parameters $\mathbf{W}$ may alleviate some of this burden.

Finally, we note that while this work does demonstrate improved log-likelihood and equivariance error, the study is inherently preliminary and does not examine all important benefits of topographic or approximately equivariant representations. Specifically, further study of the TVAE both with and without temporal coherence in terms of the sample complexity, semi-supervised classification accuracy, and invariance through structured topographic pooling would be enlightening.

3.6 Conclusion

In the above work we introduce the Topographic Variational Autoencoder as a method to train deep topographic generative models, and show how topography can be leveraged to learn approximately equivariant sets of features, a.k.a. capsules, directly from sequences of data with no other supervision. Ultimately, we believe these results may shine some light on how biological systems could hard-wire themselves to more effectively learn representations with equivariant capsule structure. In terms of broader impact, it is foreseeable our model could be used to generate more realistic transformations of 'deepfakes', enhancing disinformation. Given that the model learns *approximate* equivariance, we caution against the over-reliance on equivariant properties as these have no known formal guarantees.

3.7 Experiment Details

The code for reproducing all experiments in this chapter can be found in the following GitHub repository: https://github.com/AKAndyKeller/TopographicVAE.

3.7.1 Optimizer Parameters

Given the differences between the training procedures of the model presented in Sects. 3.4.2 and 3.4.3, the optimizer parameters for the two settings differed slightly. The 2D Topographic VAE without Temporal Coherence presented in Fig. 3.3 was trained with stochastic gradient descent on batches of size 128, using a learning rate of 1×10^{-4}, and standard momentum of 0.9 for 250 epochs. All models in Sect. 6.3 were trained with stochastic gradient descent on batches of size 8 (due to each batch-example being a length 15 or 18 sequence), using a learning rate of 1×10^{-4}, and standard momentum of 0.9 for 100 epochs.

3.7.2 Initalization

All weights of the models were initialized with uniformly random samples from $U(-\frac{1}{\sqrt{m}}, \frac{1}{\sqrt{m}})$, where m is the number of input units. For all topographic models including BubbleVAE, μ was initialized to a large value (30.0) as this was observed to increase the speed of convergence and was sometimes necessary for observed topographic organization in deeper models. For the 2D topographic model in Fig. 3.3, μ was initialized to 10.

3.7.3 Model Architectures

All models presented in this Chapter make use of the same 3-Layer MLP for parameterizing the encoders and decoders. Specifically, the model is constructed as 3 fully connected layers with ReLU activations in-between the layers. For MNIST, the layers of both the **u** and **z** encoders have (972, 648, 648) output units each for the first, second, and third layers respectively. The 648 units in the third layer are divided into two sets to compute the mean and log standard deviation of the respective u's and z's, yielding 324 t variables. This is then divided into 18 capsules, each of 18 dimensions. The layers of the decoder have (648, 972, 2352) output units respectively. For dSprites, both encoder layers have output sizes (674, 450, 450), where the resulting 225 t variables are divided into 15 capsules, each of 15 dimensions. The decoder layers then have output sizes (450, 675, 4096). We note the non-topographic VAE baselines make use of only a single encoder for the Gaussian variable **z** (as **u** is not needed), and do not incorporate a μ parameter.

3.7.4 Choices of W, W_δ, and L

Choice of W. For all topographic models (TVAE and BubbleVAE) in Sect. 3.4.3, the global topographic organization afforded by **W** was fixed to a set of 1-D tori ('circular capsules') as depicted in Fig. 3.1. The model presented in Sect. 3.4.2 organizes its variables as a single 2-D torus. Practically, multiplication by **W** was performed by convolution over the appropriate dimensions (time & capsule dimension) with a kernel of all 1's, taking advantage of circular padding to achieve toroidal structure.

Choice of W_δ. The choice of W_δ determines the local topographic structure within a single timestep. For all TVAE models with $L > 0$, we experimented with local neighborhood sizes (denoted K) of 3 units (effective kernel size 3 in the capsule dimension), and 1 unit (no neighborhood). For MNIST it was observed that $K = 3$ performed best, while $K = 1$ worked best for dSprites. This is likely due to the slower, smoother, and more overlapping transformations constructed on MNIST, whereas our subset of dSprites contained non-smooth transformations where the overlap between successive images was smaller (e.g. due to sub-setting, see Sect. 3.7.7), which made larger neighborhood sizes $K > 1$ less fitting. For TVAE models with $L = 0$, $W_\delta = W$ was fixed to sum over neighborhoods of size $K = 9$ for MNIST and $K = 3$ for dSprites. These values were chosen to be sufficiently large to achieve notably lower equivariance error than the VAE baseline, and thus demonstrate the impact of topographic organization without temporal coherence. For BubbleVAE models, the extent of topographic organization in the capsule dimension was set to $K = 3$ on MNIST to match the TVAE, and was set to be equal to the organization in time dimension $K = 2L$ for dSprites. A further quantitative comparison on the impact of the choice of the K parameter can be found in Sect. 3.8.2.

Choice of L. The choice of L determines the extent of temporal coherence where $2L$ equals the input sequence length, and $L = 0$ corresponds to single inputs. For Table 3.1, we experimented with values of L in the set $\{0, \frac{5}{36}S, \frac{1}{4}S, \frac{1}{2}S\}$ for both the TVAE and BubbleVAE. Both the BubbleVAE and TVAE achieved highest likelihoods at $L = \frac{5}{36}S$, and TVAE achieved lowest equivariance error at $L = \frac{1}{2}S$. We additionally included TVAE experiments with $L = \frac{13}{36}S$ for purposes of visualization in Figs. 3.1 and 3.4 as this yielded the best qualitative generalization. For Table 3.2, we experimented with values of L in the set

3.7 Experiment Details

$\{0, \frac{1}{6}S, \frac{4}{15}S, \frac{1}{3}S, \frac{2}{5}S, \frac{1}{2}S\}$ for both TVAE and BubbleVAE, and presented a broad selection in the table. The results of all models are shown in Sect. 3.8 below.

3.7.5 Hyperparameter Selection

Hyperparameters such as learning rate, batch size, number of capsules, capsule size, and ultimately model architecture were chosen to allow for quick training on limited resources and were not tuned significantly. Since it was conceptually simpler to have an equal number of capsule dimensions and sequence elements, this limited the number of capsules we could then train efficiently. In Sect. 3.9.1 we explain how a model with fewer capsule dimensions than sequence elements could be constructed with an alternative Roll operator. Additionally, from preliminary experiments, we observe that models with a number of internal capsule dimensions different from the number of sequence elements achieve similar likelihood values while also learning coherent transformations as decoded through the capsule roll. We believe these findings in combination with the extra studies provided in Sect. 3.8 suggest a satisfying degree of robustness to hyperparameter selection.

3.7.6 MNIST Transformations

The first set of experiments presented in this Chapter are based on the MNIST dataset [29] (MIT License). For Sect. 3.4.2 an MNIST training set of 48,000 images was used, while the standard test set of 10,000 images was used to compute the maximum activating image. For Sect. 3.4.3, sequences of MNIST images were created by picking a random training image (with a random transformation 'pose') and successively transforming it according to one of the 3 available transformations (e.g. only one attribute is changed per sequence). The available transformations consisted of rotation, color (hue rotation), and scale with increments of 20-degrees for rotation and color, and 3.66% increments for scale. Since scale is inherently non-cyclic, the bounds of the transformation were set at 60% and 126%, and the transformations were constructed to be periodic such then once scale reached 126%, the next element was at 60% scale. The final sequences were thus constructed to be 18 images long, where each element in the batch had an independently randomly chosen transformation. Again, the likelihood log $p(\mathbf{x})$ and equivariance error $\mathcal{E}_{eq}$ were computed on the held-out 10,000 example test set, where the same random transformation sequences were applied.

3.7.7 dSprites Transformations

The second set of experiments presented in this Chapter are based on the dSprites dataset [28] (Apache-2.0 License). To reduce computational complexity of this dataset, we took a subset of the dataset which consisted of all 3 shapes, the largest 5 scales, and every other example from the first 30 orientations, x-positions, and y-positions. The resulting dataset thus had 50,625 total images (3 shapes, 5 scales, 15 orientations, 15 x-positions, 15 y-positions), compared to the original 737,280 images. To construct sequences, we followed the same procedure as for MNIST, whereby first a random example and transformation were chosen, and a sequence of 15 images was constructed where only the chosen transformation was applied successively. We define the transformations available for sequences as scale, orientation, x-position, and y-position, omitting shape since smooth shape transforms are not present in the dSprites dataset. Again, we define all transformations to be cyclic such that once the 15th element is reached, the 1st element follows. For scale transformations, we simply loop over all 5 scales 3 times per sequence. We observe that although these sequences do not match the latent priors exactly, the models still train relatively well, implying some degree of robustness.

3.7.8 Capsule Correlation Metric (CapCorr)

Here we define CapCorr more precisely as it is implemented in our work. First, we denote the ground truth transformation parameter of the sequence at timestep l as y_l (e.g. the rotation angle at timestep l for a rotation sequence), and the corresponding activation at time l as $\mathbf{t}_l$. Next, to get an arbitrary starting point, we let $l = \Omega$ denote the timestep when y_l is at its canonical position (e.g. rotation angle 0, x-position 0, or scale 1). We see Ω is not necessarily 0 since the first timestep of each sequence ($l = 0$) is a randomly transformed example. Then, we observe that we can measure the approximate observed roll in the capsule dimension between time 0 and Ω as a 'phase shift' by computing the index of the maximum value of a discrete (periodic) cross-correlation of $\mathbf{t}_\Omega$ and $\mathbf{t}_0$:

$$\text{ObservedRoll}(\mathbf{t}_\Omega, \mathbf{t}_0) = \text{argmax}\,[\mathbf{t}_\Omega \star \mathbf{t}_0] \qquad (3.15)$$

where $\star$ is discrete (periodic) cross-correlation across the (cyclic) capsule dimension and argmax is also subsequently performed over the capsule dimension. Then, the CapCorr metric for a single capsule is given as:

$$\text{CapCorr}(\mathbf{t}_\Omega, \mathbf{t}_0, y_\Omega, y_0) = \text{Corr}\,(\text{ObservedRoll}(\mathbf{t}_\Omega, \mathbf{t}_0), |y_\Omega - y_0|) \qquad (3.16)$$

where the correlation coefficient Corr is then computed across all examples for the entire dataset. In our experiments we use the Pearson correlation coefficient for Corr. We thus see this metric is the correlation of the estimated observed capsule roll with the shift in ground truth generative factors, which is equal to 1 when the model is perfectly equivariant. To extend this definition to multiple capsules, we estimate ObservedRoll for each capsule separately, and then correlate the mode of all ObservedRoll values with the true shift in ground truth generative factors. We see empirically that the ObservedRolls for all capsules are almost

3.8 Extended Results

Table 3.3 Log Likelihood and Equivariance Error on MNIST for all models tested. Mean ± std. over 3 random initalizations

Model	TVAE	TVAE	TVAE	TVAE	TVAE
L	$L = \frac{1}{2}S$	$L = \frac{13}{36}S$	$L = \frac{1}{4}S$	$L = \frac{5}{36}S$	$L = 0$
K	$K = 3$	$K = 3$	$K = 3$	$K = 3$	$K = 9$
$\log p(\mathbf{x}) \uparrow$	$-\mathbf{186.8 \pm 0.1}$	-188.0 ± 0.5	-187.0 ± 0.2	-186.0 ± 0.7	-218.5 ± 0.9
$\mathcal{E}_{eq} \downarrow$	$\mathbf{573.9 \pm 1.5}$	1089.8 ± 2.4	2136.9 ± 7.8	3246.6 ± 3.3	3216.6 ± 104.9
Model	BubbleVAE	BubbleVAE	BubbleVAE	BubbleVAE	VAE
L	$L = \frac{1}{2}S$	$L = \frac{1}{4}S$	$L = \frac{5}{36}S$	$L = \frac{5}{36}S$	$L = 0$
K	$K = 2L$	$K = 2L$	$K = 2L$	$K = 3$	$K = 1$
$\log p(\mathbf{x}) \uparrow$	-200.9 ± 0.7	-202.3 ± 1.4	-190.8 ± 0.7	-191.4 ± 0.5	-189.0 ± 0.8
$\mathcal{E}_{eq} \downarrow$	4206.7 ± 903.3	1141.7 ± 9.6	2605.7 ± 16.1	3369.5 ± 11.9	13273.9 ± 0.5

always identical (i.e. all capsules roll simultaneously for each transformation), therefore computing the mode does not destroy significant information. Finally, for transformation sequences which have multiple timesteps where y_l is at the canonical position (e.g. scale transformations on dSprites where scale is looped 3 times), we select $l = \Omega$ to be the one from this possible set which yields the minimal absolute distance between $|y_\Omega - y_0|$ and ObservedRoll($\mathbf{t}_\Omega, \mathbf{t}_0$).

3.7.9 Definition of Roll for Capsules

As stated in Sect. 4.5.2, $\text{Roll}_\delta(\mathbf{u})$, is defined as a cyclic permutation of δ steps along the capsule dimension of $\mathbf{u}$. Explicitly, if $\mathbf{u}$ is divided into C capsules each with D dimensions, the Roll_δ operation can be written as:

$$\text{Roll}_\delta(\mathbf{u}) = \text{Roll}_\delta([u_1, u_2, \ldots, u_{C \cdot D}])$$
$$= [u_D, u_1, \ldots, u_{D-1}, u_{2 \cdot D}, u_{D+1}, \ldots, u_{2 \cdot D-1}, u_{3 \cdot D}, \ldots, \ldots u_{C \cdot D-1}] \quad (3.17)$$

3.8 Extended Results

In this section we provide extended results for all tested hyperparamters (Tables 3.3 and 3.4), a further analysis of the impact of the coherence window within a capsule $\mathbf{W}_\delta$ (Table 3.5), samples from the model in Sect. 3.4.2, and additional capsule traversal experiments highlighting the generalization capabilities of the TVAE to combinations of transformations unseen during training (Fig. 3.5).

Table 3.4 Equivariance error and CapCorr for all models tested on the dSprites dataset. Mean ± standard deviation over 3 random initalizations.

Model	TVAE	TVAE	TVAE	TVAE	TVAE	TVAE
L	$L = \frac{1}{2}S$	$L = \frac{2}{5}S$	$L = \frac{1}{3}S$	$L = \frac{4}{15}S$	$L = \frac{1}{6}S$	$L = 0$
K	$K = 1$	$K = 1$	$K = 1$	$K = 1$	$K = 1$	$K = 3$
CapCorr$_X$ ↑	1.0 ± 0	1.0 ± 0	1.0 ± 0	0.95 ± 0.00	0.67 ± 0.02	0.17 ± 0.03
CapCorr$_Y$ ↑	1.0 ± 0	1.0 ± 0	1.0 ± 0	0.96 ± 0.01	0.66 ± 0.02	0.21 ± 0.02
CapCorr$_O$ ↑	1.0 ± 0	1.0 ± 0	1.0 ± 0	0.88 ± 0.01	0.52 ± 0.01	0.09 ± 0.01
CapCorr$_S$ ↑	1.0 ± 0	1.0 ± 0	1.0 ± 0	0.96 ± 0.01	0.42 ± 0.01	0.51 ± 0.01
$\mathcal{E}_{eq}$ ↓	344 ± 5	759 ± 9	1034 ± 6	1395 ± 7	2549 ± 38	2971 ± 9
Model	BubbleVAE	BubbleVAE	BubbleVAE	BubbleVAE	BubbleVAE	VAE
L	$L = \frac{1}{2}S$	$L = \frac{2}{5}S$	$L = \frac{1}{3}S$	$L = \frac{4}{15}S$	$L = \frac{1}{6}S$	$L = 0$
K	$K = 2L$	$K = 2L$	$K = 2L$	$K = 2L$	$K = 2L$	$K = 1$
CapCorr$_X$ ↑	0.16 ± 0.01	0.15 ± 0.01	0.13 ± 0.01	0.12 ± 0.02	0.09 ± 0.01	0.18 ± 0.01
CapCorr$_Y$ ↑	0.15 ± 0.01	0.14 ± 0.01	0.12 ± 0.01	0.12 ± 0.01	0.11 ± 0.02	0.16 ± 0.01
CapCorr$_O$ ↑	0.12 ± 0.00	0.13 ± 0.02	0.10 ± 0.01	0.09 ± 0.00	0.06 ± 0.01	0.11 ± 0.00
CapCorr$_S$ ↑	0.52 ± 0.02	0.55 ± 0.00	0.52 ± 0.00	0.48 ± 0.02	0.27 ± 0.01	0.52 ± 0.00
$\mathcal{E}_{eq}$ ↓	6825 ± 126	6917 ± 13	1951 ± 34	2181 ± 627	1721 ± 27	6934 ± 0

Table 3.5 Impact of $\mathbf{W}_\delta$ (i.e. K) on MNIST performance

Model	TVAE	TVAE	TVAE	TVAE	TVAE
L	$L = \frac{5}{36}S$	$L = \frac{5}{36}S$	$L = 0$	$L = 0$	$L = 0$
K	$K = 3$	$K = 9$	$K = 3$	$K = 9$	$K = 18$
$\log p(\mathbf{x})$ ↑	-186.0 ± 0.7	-190.6 ± 0.2	-213.4 ± 1.2	-218.5 ± 0.9	-224.8 ± 1.0
$\mathcal{E}_{eq}$ ↓	3246.6 ± 3.3	2606.3 ± 17.0	12085.7 ± 68.5	3216.6 ± 104.9	1090.3 ± 19.3

3.8.1 Extended Tables 3.1 and 3.2

In Tables 3.3 and 3.4 below, we present extended versions of Tables 3.1 and 3.2 respectively, showing all tested settings of the TVAE & BubbleVAE. We observe the TVAE achieves perfect correlation (CapCorr = 1) for $L \geq \frac{1}{3}$, and steadily decreasing correlation for lower values of L.

3.8 Extended Results

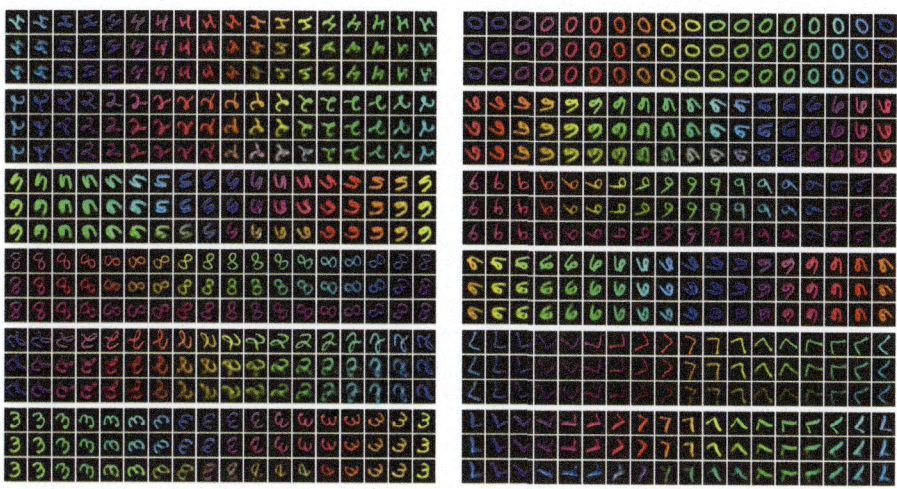

Fig. 3.5 Capsule Traversals for MNIST TVAE $L = \frac{13}{36}S$, trained on individual transformations in isolation, and tested on combined color and rotation transformations. Top row shows the input sequence, middle row shows the direct reconstruction $\{g_\theta(\mathbf{t}_l)\}_l$, and bottom row shows the capsule traversal $\{g_\theta(\text{Roll}_l[\mathbf{t}_0])\}_l$

3.8.2 Impact of $\mathbf{W}_\delta$

In Table 3.5, we show a small set of experiments with different settings of $\mathbf{W}_\delta$, and specifically changing values of K (the coherence window within a capsule). As can be seen, increasing K generally reduces equivariance error, but decreases the log-likelihood. This can be further understood by examining the capsule traversals of such models in Figs. 3.8, 3.9, 3.10, 3.11, and 3.12. We see that larger values of K appear to induce smoother transformations within the capsule dimensions, eventually resulting in invariant representations when K is equal to the capsule dimensionality (Figs. 3.6 and 3.7).

3.8.3 Generalization to Combined Transformations at Test Time

In this section, we test the ability of the model to generate sequences composed of multiple transformations through a capsule roll, despite only being trained on individual transformations in isolation. In other words, we intend to measure the extent to which the transformations learned by a set of capsules can be combined simply by passing input sequences with corresponding combined transformations. Such generalization suggests powerful benefits to data efficiency, effectively factorizing a set of complex transformations.

Explicitly, we train the model identically to that presented in Fig. 3.4, (TVAE $L = \frac{13}{36}S$), and examine the sequences generated by a capsule roll when the partial input sequences

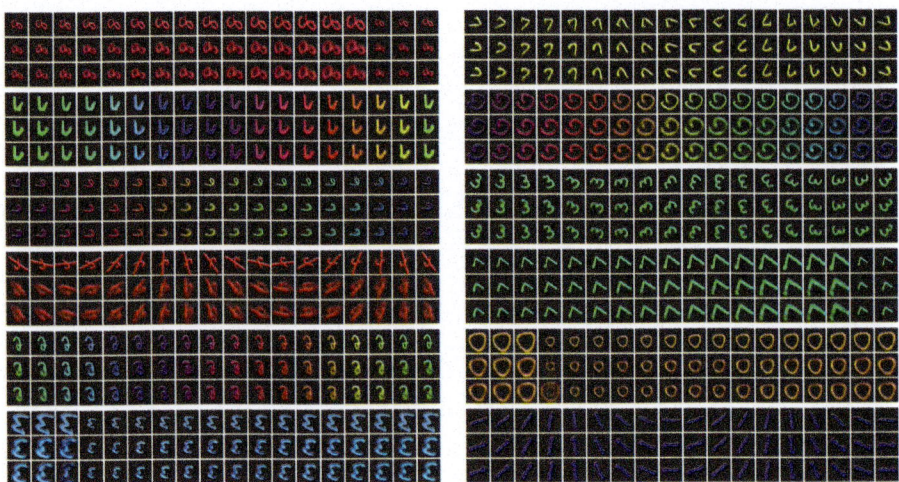

Fig. 3.6 MNIST TVAE $L = \frac{1}{2}S$, $K = 3$

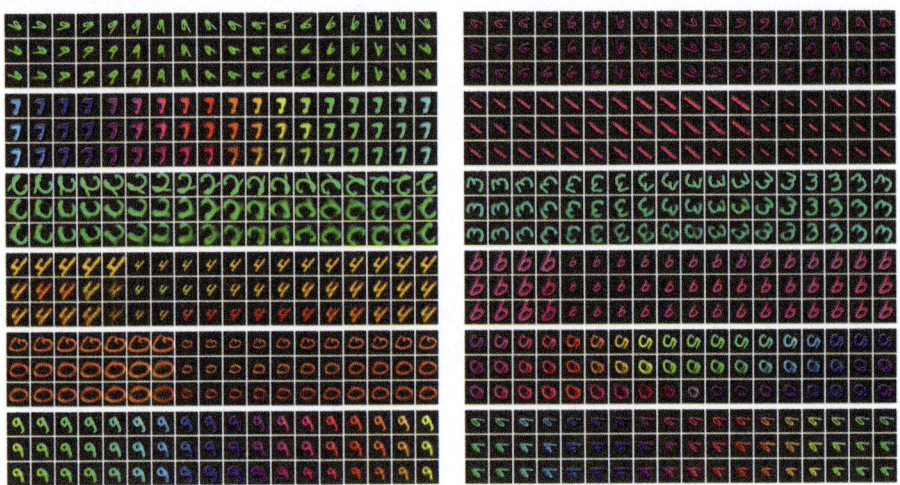

Fig. 3.7 MNIST TVAE $L = \frac{13}{36}S$, $K = 3$

contain combinations of transformations previously unseen during training. The results of this experiment, tested on combinations of rotation and color transforms on the MNIST test set, are presented in Fig. 3.5. Although this generalization capability is not known to be guaranteed a priori, we see that the capsule traversals are frequently remarkably coherent with the input transformation, implying that the model may indeed be able to generalize to combinations of transformations. Furthermore, we observe with $L = \frac{1}{2}S$ (results not shown), this generalization capability is nearly perfect.

3.9 Proposed Model Extensions

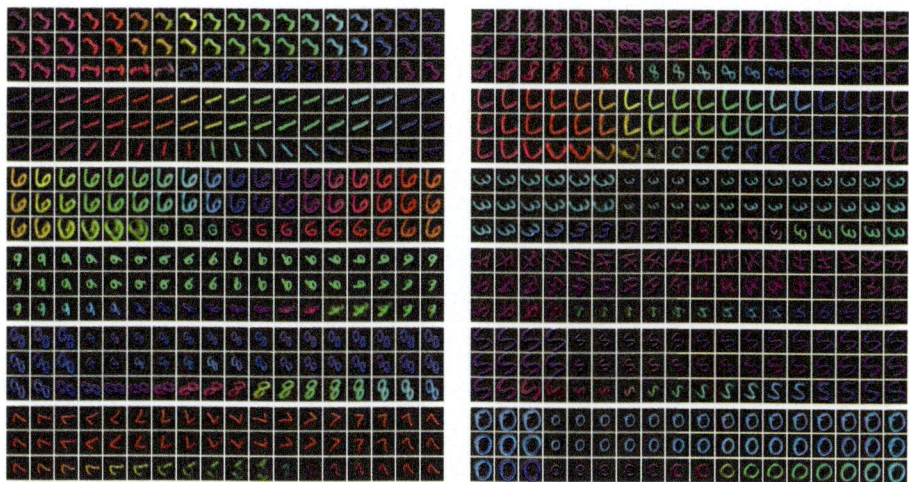

Fig. 3.8 MNIST TVAE $L = \frac{5}{36}S, K = 3$. We see with values of $L < \frac{1}{3}S$ the transformations decoded through the capsule roll are only partially coherent with the input sequence

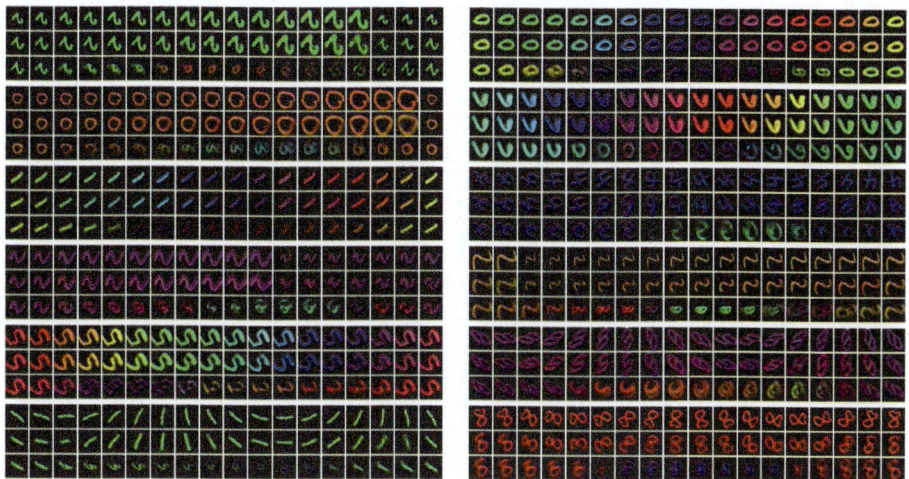

Fig. 3.9 MNIST TVAE $L = \frac{5}{36}S, K = 9$

3.9 Proposed Model Extensions

3.9.1 Extensions to Roll and CapCorr

The Roll operation can be seen as defining the speed at which **t** transforms corresponding to an observed transformation. For example, with Roll defined as in Sect. 3.7.9 above, we implicitly assume that for each observed timestep, we would like the representation **t** to

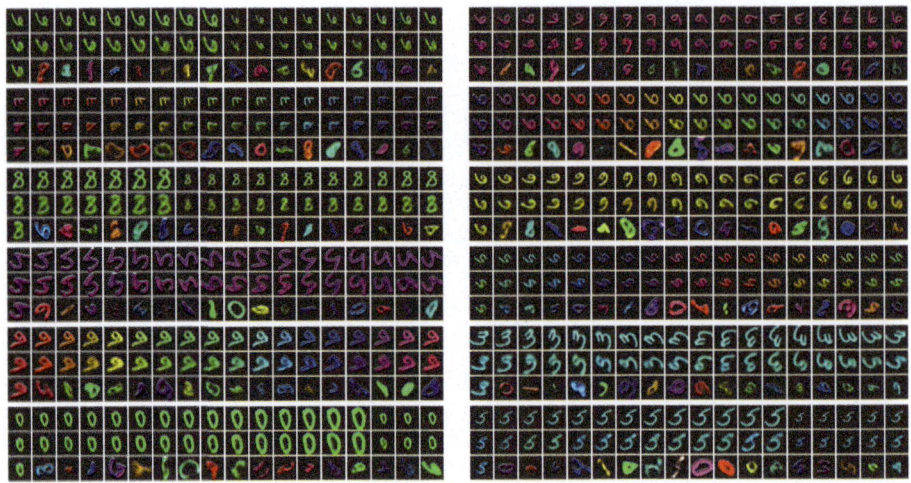

Fig. 3.10 MNIST TVAE $L = 0$, $K = 3$. We see for sufficiently small values of K, the TVAE can reach a degenerate solution where topographic organization is almost entirely lost

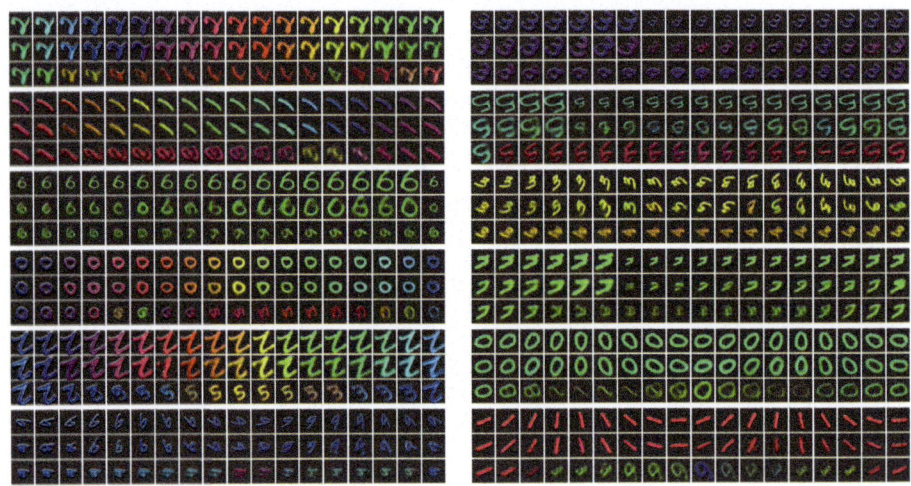

Fig. 3.11 MNIST TVAE $L = 0$, $K = 9$

cyclically permute 1-unit within the capsule. For this to match the observed data, it requires the model to have an equal number of capsule dimensions and sequence elements. If we wish to reduce the size of our representation, we could instead encourage a 'partial permutation' for each observed transformation. For a single capsule with D elements, an example of a simple linear version of such a partial permutation (for $0 < \alpha \leq 1$) can be implemented as:

$$\text{Roll}_\alpha(\mathbf{u}) = \begin{bmatrix} \alpha u_D + (1-\alpha)u_1, & \alpha u_1 + (1-\alpha)u_2, & \ldots, & \alpha u_{D-1} + (1-\alpha)u_D \end{bmatrix} \quad (3.18)$$

3.9 Proposed Model Extensions

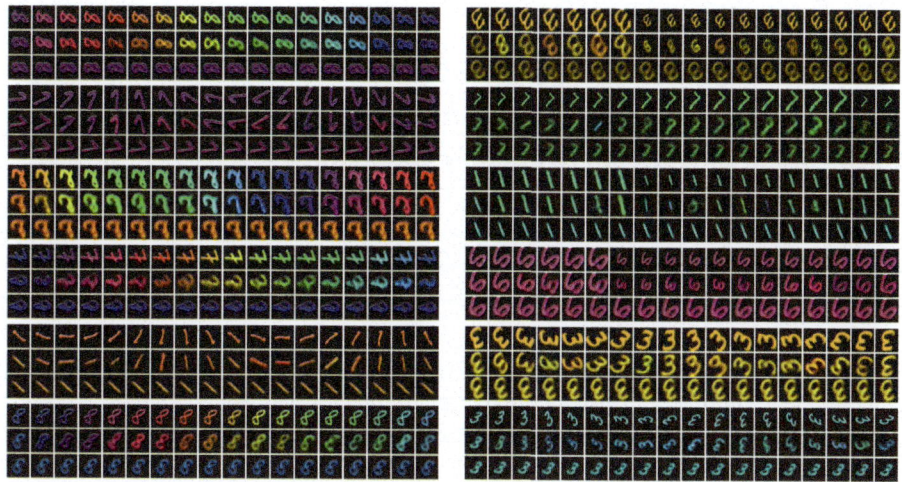

Fig. 3.12 MNIST TVAE $L = 0$, $K = 18$. We see when K is equal to the capsule size (making the model analogous to ISA), the model learns an invariant capsule representation – meaning Rolling a capsule activation produces no significant transformation in the observation space

A slightly more principled partial roll for periodic signals could also be achieved by performing a phase shift of the signal in Fourier space, and performing the inverse Fourier transform to obtain the resulting rolled signal. To extend the CapCorr metric to similarly allow for partial Rolls, we see that we can simply redefine the ObservedRoll (originally given by discrete cross-correlation) to be given by the argmax of the inner product of a sequentially partially rolled activation with the initial activation $\mathbf{t}_\Omega$. Formally:

$$\text{ObservedRoll}(\mathbf{t}_\Omega, \mathbf{t}_0) = \text{argmax}\Big[\mathbf{t}_\Omega \cdot \text{Roll}_0(\mathbf{t}_0),$$
$$\mathbf{t}_\Omega \cdot \text{Roll}_\alpha(\mathbf{t}_0), \ldots, \mathbf{t}_\Omega \cdot \text{Roll}_{D-\alpha}(\mathbf{t}_0)\Big] \qquad (3.19)$$

3.9.2 Non-cyclic Capsules

We can also see that there is nothing beyond convenience which inherently requires the capsules to be circular (i.e. have periodic boundary conditions). To implement linear capsules, we propose one solution is to add L additional U_i variables to both the left and right boundaries of each capsule. In this way, the vector $\mathbf{U}$ is larger than the vector $\mathbf{Z}$ and can be seen as a 'padded' version, where the padding is composed of independant random variables. Additionally, the transformation sequences can then be padded on both sides by replicating the first and final elements L times. The construction of $\mathbf{T}$ variables is then performed identically as in Eqs. 3.7 and 3.9. The Roll operation can then be similarly defined as filling the boundaries with 0 since these values will not be used as part of the computation.

3.9.3 Multi-dimensional Temporally Coherent Capsules

In consideration of transformations which may naturally live in multiple dimensions, we wish to extend the original model to support multi-dimensional capsules. Such multi-dimensional capsules could additionally support more well-defined 'disentanglement' of transformations by encouraging each transformation to be axis-aligned with one dimension of each capsule. We see that in the non-temporally coherent case ($L = 0$), the model can easily be extended to capsules of multiple dimensions through multi-dimensional neighborhoods. An example of a model with 2-dimensional neighborhoods is presented in Fig. 3.3. However, when considering shifting temporal coherence as we defined in Sect. 3.4.3, it is not clear how the shift operator or the neighborhoods should be defined for higher dimensional capsules. In this section we propose to modify the definitions of $\mathbf{T}$ in Eqs. 3.7 and 3.9 with an extension resembling 'group sparsity' in the denominator.

First, we again assume that each input sequence is an observation of a single transformation at a time. Formally, the multi-dimensional capsules are then constructed by arranging $\mathbf{U}$ into a D dimensional lattice. In such a model, we desire to roll and sum only along a single axis of the lattice for a given sequence. Incorporating this into the construction of $\mathbf{T}$ yields the following:

$$\mathbf{T}_l = \frac{\mathbf{Z}_l - \mu}{\sum_{d=1}^{D} \sqrt{\mathbf{W}^d \left[\mathbf{U}_{l+L}^2 ; \cdots ; \mathbf{U}_{l-L}^2 \right]}} = \frac{\mathbf{Z}_l - \mu}{\sum_{d=1}^{D} \sqrt{\sum_{\delta=-L}^{L} \mathbf{W}_\delta^d \mathrm{Roll}_\delta^d (\mathbf{U}_{l+\delta}^2)}} \qquad (3.20)$$

where $\mathbf{W}_\delta^d$ refers to a matrix which sums locally along the dth dimension of each capsule, and not at all along the others, and similarly Roll_δ^d rolls only along the dth dimension. In practice we observe such models can indeed disentangle up to 2 distinct transformations, but become more challenging to optimize for higher dimensions. We believe this is potentially due to the exponential growth in capsule size with increasing dimension, but leave further exploration to future work.

3.9.4 Causal Temporal Coherence

As noted in the limitations, the sequence model is not 'causal', meaning that each variable $\mathbf{T}_l$ requires variables from future timesteps in the sequence ($\mathbf{U}_{l+\delta}$ for $\delta > 0$). Although for the purpose of learning equivariance in practice this may not be an issue, it may be relevant for some online learning applications. We can modify Eqs. 3.7 and 3.9 by changing the matrix $\mathbf{W}$ (implemented as convolution) to a causal convolution (i.e. masking out $\mathbf{W}_\delta$ for $\delta > 0$).

3.10 Capsule Traversals

Formally:

$$T_l = \frac{Z_l - \mu}{\sqrt{W\left[U_l^2; \cdots ; U_{l-L}^2\right]}} = \frac{Z_l - \mu}{\sqrt{\sum_{\delta=-L}^{0} W_\delta \text{Roll}_\delta(U_{l+\delta}^2)}} \quad (3.21)$$

In a causal setting, it is also likely the transformations are no longer assumed to be circular. We thus refer the reader to Sect. 3.9.2 above on non-circular capsules, which can be combined with Eq. 3.21, to achieve such a model.

3.10 Capsule Traversals

In this section we provide an additional set of capsule traversals to compliment those presented presented in main text. Unlike the main section, we additionally include a middle row which shows the direct reconstruction of the input without any rolling (i.e. $\{g_\theta(t_l)\}_l$). We find the direct reconstructions valuable to determine if poor traversals are due to bad reconstructions (low $\log p_\theta(x|t)$) or a lack of equivariance (high $\mathcal{E}_{eq}$). For example, with the baseline VAE models, we see that the reconstructions in the middle row are accurate for the full sequence, while the capsule traversals obtained by sequentially rolling the initial activation (shown in the bottom row) are nothing like the input transformation (top row). In all traversals, the left-most image corresponds to t_0, and thus input sequences of length $2L$ cover both the left and right edges when $L > 0$ (Fig. 3.13).

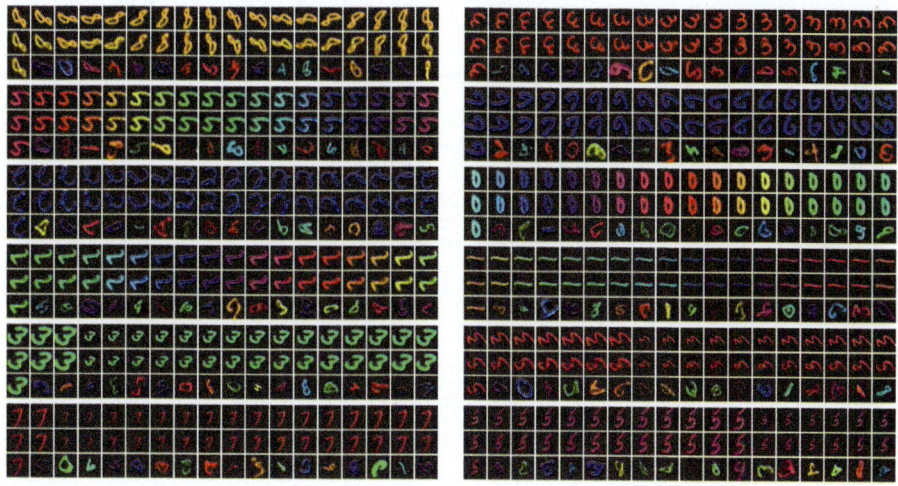

Fig. 3.13 MNIST VAE $L = 0$, $K = 1$. We see images generated through capsule traversal with the baseline VAE appear entirely random, as expected due to the non-topographic nature of the VAE's latent space

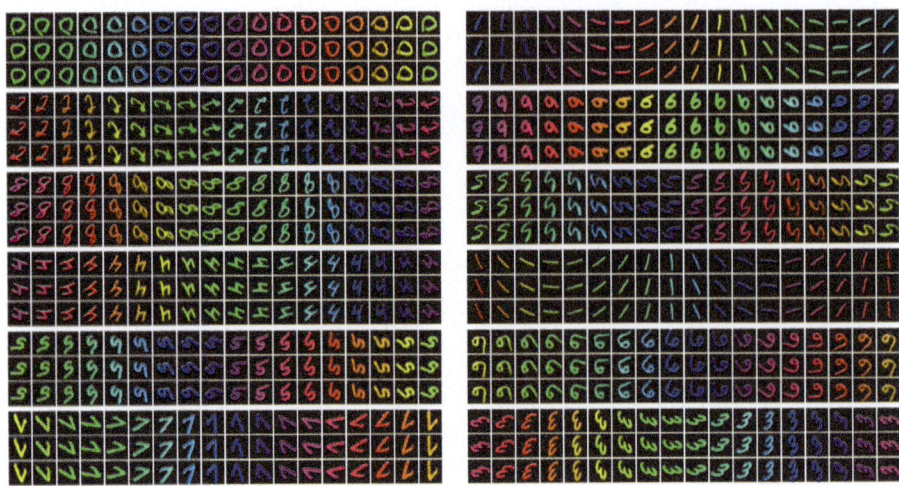

Fig. 3.14 Combined Color & Rotation MNIST TVAE $L = \frac{13}{36}S$, $K = 3$. We see these generated sequences are slightly more accurate than those in Fig. 3.5. This is to be expected since the model in this figure is trained explicitly on combinations of transformations, whereas the model in Fig. 3.5 was trained on transformations in isolation, and tested on combinations to explore its generalization

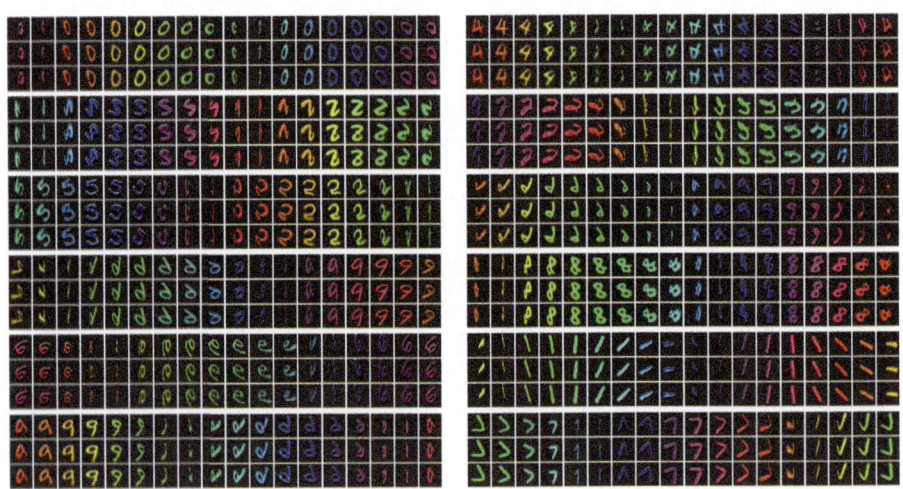

Fig. 3.15 Combined Color & Perspective MNIST TVAE $L = \frac{13}{36}S$, $K = 3$. We see the TVAE is able to additionally learn combinations of complex transformations (like out-of-plane rotation) without any changes to the training procedure other than a change of dataset

Finally, in Figs. 3.14 and 3.15 at the end of the section, we include capsule traversals for models trained on MNIST with more complex transformations such as combined color &

rotation, and combined color & perspective transforms. These models were trained in an identical manner to the other MNIST models, with the same architecture, only changing the transformation sequences of the training dataset.

References

1. Horace B Barlow et al. Possible principles underlying the transformation of sensory messages. *Sensory communication*, 1(01), 1961.
2. Anthony J. Bell and Terrence J. Sejnowski. An Information-Maximization Approach to Blind Separation and Blind Deconvolution. *Neural Computation*, 7(6):1129–1159, 11 1995.
3. Pierre Comon. Independent component analysis, a new concept? *Signal processing*, 36(3):287–314, 1994.
4. Aapo Hyvärinen and Erkki Oja. Independent component analysis: algorithms and applications. *Neural networks*, 13(4-5):411–430, 2000.
5. Bruno A Olshausen and David J Field. Sparse coding with an overcomplete basis set: A strategy employed by V1? *Vision research*, 37(23):3311–3325, 1997.
6. J Portilla, V Strela, M J Wainwright, and E P Simoncelli. Image denoising using scale mixtures of Gaussians in the wavelet domain. *IEEE Trans Image Processing*, 12(11):1338–1351, Nov 2003. Recipient, IEEE Signal Processing Society Best Paper Award, 2008.
7. M J Wainwright and E P Simoncelli. Scale mixtures of Gaussians and the statistics of natural images. In S. A. Solla, T. K. Leen, and K.-R. Müller, editors, *Adv. Neural Information Processing Systems (NIPS*99)*, volume 12, pages 855–861, Cambridge, MA, 2000. MIT Press.
8. Aapo Hyvärinen, Patrik O Hoyer, and Mika Inki. Topographic independent component analysis. *Neural computation*, 13(7):1527–1558, 2001.
9. Simon Osindero, Max Welling, and Geoffrey E. Hinton. Topographic Product Models Applied to Natural Scene Statistics. *Neural Computation*, 18(2):381–414, 02 2006.
10. Simon Kayode Osindero. *Contrastive Topographic Models*. PhD thesis, University of London, 2004.
11. Max Welling, Simon Osindero, and Geoffrey E Hinton. Learning sparse topographic representations with products of student-t distributions. In *Advances in neural information processing systems*, pages 1383–1390, 2003.
12. Aapo Hyvärinen, Jarmo Hurri, and Patrick O Hoyer. *Natural image statistics: A probabilistic approach to early computational vision.*, volume 39. Springer Science & Business Media, 2009.
13. Aapo Hyvärinen and Patrik O. Hoyer. A two-layer sparse coding model learns simple and complex cell receptive fields and topography from natural images. *Vision Research*, 41(18):2413–2423, 2001.
14. Libo Ma and Liqing Zhang. Overcomplete topographic independent component analysis. *Neurocomputing*, 71(10-12):2217–2223, 2008.
15. S Lyu and E P Simoncelli. Nonlinear image representation using divisive normalization. In *Proc. Computer Vision and Pattern Recognition*, pages 1–8. IEEE Computer Society, Jun 23-28 2008.
16. S Lyu and E P Simoncelli. Modeling multiscale subbands of photographic images with fields of Gaussian scale mixtures. *IEEE Trans. Patt. Analysis and Machine Intelligence*, 31(4):693–706, Apr 2009.
17. M J Wainwright, E P Simoncelli, and A S Willsky. Random cascades on wavelet trees and their use in analyzing and modeling natural images. *Applied and Computational Harmonic Analysis*, 11(1):89–123, 2001.

18. Gregory Benton, Marc Finzi, Pavel Izmailov, and Andrew Gordon Wilson. Learning invariances in neural networks. *Advances in Neural Information Processing Systems*, December, 2020.
19. Nichita Diaconu and Daniel Worrall. Learning to convolve: A generalized weight-tying approach. In Kamalika Chaudhuri and Ruslan Salakhutdinov, editors, *Proceedings of the 36th International Conference on Machine Learning*, volume 97 of *Proceedings of Machine Learning Research*, pages 1586–1595. PMLR, 09–15 Jun 2019.
20. Peter Földiák. Learning invariance from transformation sequences. *Neural Computation*, 3:194–200, 06 1991.
21. Jarmo Hurri and Aapo Hyvärinen. Simple-Cell-Like Receptive Fields Maximize Temporal Coherence in Natural Video. *Neural Computation*, 15(3):663–691, 03 2003.
22. James V. Stone. Learning Perceptually Salient Visual Parameters Using Spatiotemporal Smoothness Constraints. *Neural Computation*, 8(7):1463–1492, 10 1996.
23. Laurenz Wiskott and Terrence J Sejnowski. Slow feature analysis: Unsupervised learning of invariances. *Neural computation*, 14(4):715–770, 2002.
24. Aapo Hyvärinen and Patrik Hoyer. Emergence of phase-and shift-invariant features by decomposition of natural images into independent feature subspaces. *Neural computation*, 12(7):1705–1720, 2000.
25. A. Hyvärinen, J. Hurri, and Jaakko J. Väyrynen. A unifying framework for natural image statistics: spatiotemporal activity bubbles. *Neurocomputing*, 58-60:801–806, 2004.
26. Geoffrey E. Hinton and Yee-Whye Teh. Discovering multiple constraints that are frequently approximately satisfied. In *Proceedings of the Seventeenth Conference on Uncertainty in Artificial Intelligence*, UAI'01, page 227-234, 2001.
27. Teuvo Kohonen. Emergence of invariant-feature detectors in the adaptive-subspace self-organizing map. *Biological cybernetics*, 75(4):281–291, 1996.
28. Loic Matthey, Irina Higgins, Demis Hassabis, and Alexander Lerchner. dsprites: Disentanglement testing sprites dataset, 2017.
29. Yann LeCun, Corinna Cortes, and CJ Burges. Mnist handwritten digit database. *ATT Labs [Online]. Available:* http://yann.lecun.com/exdb/mnist, 2, 2010.

Neural Wave Machines

4.1 Introduction

In machine learning, inductive biases can be understood as limiting the search space of possible hypotheses a priori, and indeed, it is known that without any inductive bias, learning generalizations beyond the training data is theoretically impossible [1]. Modern machine learning researchers have adopted many task-specific inductive biases almost by default, such as convolution for spatially structured data. Similarly, natural intelligence as implemented by biological systems also has many inductive biases by virtue of the diversity of constraints that it must simultaneously satisfy such as metabolic efficiency. The fields of psychology, cognitive science, and neuroscience have all studied these biases and their observed signatures, often hypothesizing about their computational implications (Fig. 4.1).

One such observation which has recently gained increasing interest in the neuroscience community is that of traveling waves of neural activity. Such waves have been measured at both local [2] and global [3] scales, and have been shown to be strongly related to alpha, theta, and gamma oscillations in a variety of brain regions [4, 5]. Prompted by these observations, a large number of theoretical hypotheses have been developed which attempt to explain the computational purposes of traveling waves [6], and the inductive biases which they may mediate.

Of particular relevance to the machine learning community, one hypothesis is that traveling waves serve to beneficially structure neural representations in both space and time [7, 8], acting as an inductive bias towards similarly structured natural data. Structured representations have been previously demonstrated in the machine learning community to be extremely valuable, making learning models both more efficient and robust [9]. A prime example of this is group equivariance [10]; in the case of translation, this resulted in the convolutional neural network which reduced the sensitivity of existing fully connected artificial neural networks to small image shifts and deformations [11, 12], thereby facilitating the rapid growth of the field of deep learning [13]. In the case of traveling waves, it is thus

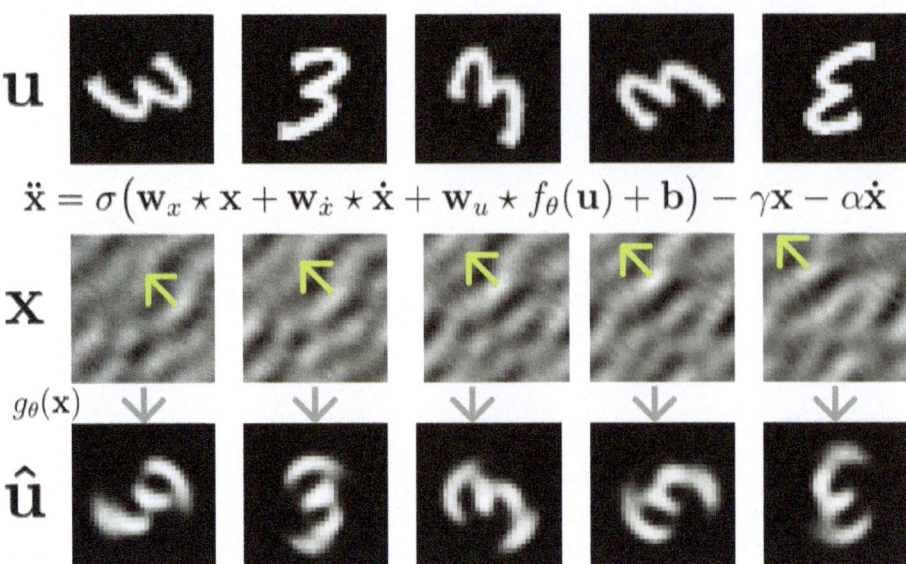

Fig. 4.1 Overview of the Neural Wave Machine. The input sequence **u** is encoded with f_θ to act as a driving term in the hidden state **x** which is modeled temporally ($\ddot{\mathbf{x}}$) as a network of locally coupled oscillators. The network is then trained to reconstruct the input sequence: $\hat{\mathbf{u}} = g_\theta(\mathbf{x})$. The yellow arrows track a traveling wavefront over time

suggested that they may facilitate a similar kind of spatiotemporal structure in neural representations, thereby granting the observed robustness and efficiency of natural intelligence which is still lacking in modern deep neural networks [14].

To date, however, testing ideas related to the computational purposes of traveling waves has been challenging due to a lack of neural network architectures which have a notion of spatial locality necessary for modeling such spatio-temporal dynamics. Further, existing networks which do have such spatial structure often do not have temporal structure [15, 16], or are not sufficiently flexibly parameterized to allow them to be trained on standard machine learning benchmarks [17].

In this work, we propose to investigate the computational hypotheses surrounding traveling waves through a bottom-up approach; we build a flexibly parameterized computational model known to be capable of producing traveling waves, and show that it indeed learns to exhibit complex spatiotemporal dynamics when modeling real data. We then show, relevant to the computational neuroscience community, how such a network indeed learns spatial and temporal structure reminiscent of that found in the brain. Specifically, we observe that our network learns topographically organized selectivity, similar to the observed orientation columns and hypercolumns of the primary visual cortex [18]. Further, we show that our network learns to use complex spatiotemporal organization such as traveling waves to encode

4.1 Introduction

transformations by artificially inducing waves in the hidden state and observing that this allows us to further progress or reverse the transformations of generated images.

As it relates to inductive biases, we asses the computational implications of the observed representational structure by training the model on the physical dynamics forecasting suite introduced in the paper 'Which Priors Matter?' [19]. We see that our model is more accurate at predicting future trajectories of simple physical dynamics when compared with existing state of the art models, providing evidence that the structure mediated by traveling waves is indeed a beneficial inductive bias for modeling such smooth natural transformations. Further, due to our model's local connectivity, we see that it is more efficient both in terms of parameters, and in terms of biological concerns such as wiring length, suggesting a connection between locality of connections, waves, and an inductive transformation bias in biological systems.

Overall we believe our work offers the concrete contribution of a new powerful model at the interface of computational neuroscience and modern machine learning. We show that this model allows for the investigation of the computational hypotheses surrounding complex synchrony in the brain in a new way, and further provides preliminary evidence for the existing hypothesis that traveling waves serve to induce spatiotemporal structure in neural representations.

4.1.1 Traveling Waves in Neuroscience

Neural oscillations and traveling waves have long been a subject of study in neuroscience and neurophysiology [6, 20]. Although such waves were originally measured primarily in anesthetized subjects, improved multi-channel recording and analysis techniques have recently demonstrated propagating wave activity in awake functioning subjects as well, originating from both external stimuli and internal 'spontaneous' recurrent connections [6, 21, 22]. While many hypotheses have been put forth for their precise computational role, a consensus has yet to be reached. Example hypotheses include that traveling waves may: influence visual perception [23]; modulate information transfer [5]; correlate with conscious awareness [24]; facilitate predictive coding [25, 26]; lower the threshold for detection of weak stimuli [2]; serve as a short term memory [27, 28]; or as a mechanism for the formation of long-term memories during sleep [29]. Relevant to this work, traveling waves have directly been implicated in the encoding of motion [30], and have been measured to correlate strongly with perceived perceptual illusions of motion [8]. Further, it has been suggested that they form the basis of alpha and theta oscillations [4, 7] and may serve to both structure and integrate information across space and time [21, 31]. Due to the fundamental relationship between neural synchrony and the coordination of spike timing [32], it is natural to wonder if more complex forms of spatiotemporal synchrony such as traveling waves may play a similarly more complex structural role.

4.1.2 Computational Models of Traveling Waves

In the fields of computational and theoretical neuroscience, multiple models have been developed to help explain the observed complex synchronous dynamics of neural systems. One classical model is that of a network of locally coupled oscillators [33, 34]. However, to date, such models have been limited to those which either are built for the primary purpose of analysis [17, 35, 36], or those which perform very simple binary operations [37, 38], with neither set leveraging the flexible computational capabilities of modern deep neural networks. One line of work has aimed to integrate classical Kuramoto models into deep neural networks by directly parameterizing activations in terms of phase values [39], however such models lack a notion of spatial locality, making the existence of spatio-temporal dynamics less concrete. Most recently, Davis et al. [17] studied a large scale locally connected spiking neural network model, quantifying the conditions necessary for the emergence of traveling waves, and showed such waves appeared to uniquely agree with human cortical traveling waves in a variety of dimensions. However, similar to most existing models in this category, the model is formulated as a spiking neural network thus requiring more sophisticated training mechanisms which are yet to scale to the same performance as deep neural networks [40].

4.2 Neural Wave Machines

In the following section we introduce the Neural Wave Machine (NWM), a deep neural network architecture which exhibits traveling waves and other complex spatiotemporal dynamics in the service of flexible differentiable computation. To achieve this, we take inspiration from the seminal models of traveling waves built as networks of locally coupled oscillators [36], and propose to integrate them into a modern deep learning framework by taking advantage of the recently developed coupled oscillatory Recurrent Neural Network (coRNN) [41].

4.2.1 Coupled Oscillatory Recurrent Neural Networks

In [41] the authors propose to solve the Exploding and Vanishing Gradient Problem (EVGP) in recurrent neural networks by defining a new recurrent neural network with hidden state dynamics given by the parameterized equations of a system of coupled, damped, and driven oscillators. Explicitly, the hidden state of the recurrent neural network $\mathbf{x}$ is updated by solving the following second order partial differential equation:

$$\ddot{\mathbf{x}} = \sigma \left(\mathbf{W}_x \mathbf{x} + \mathbf{W}_{\dot{x}} \dot{\mathbf{x}} + \mathbf{V} \mathbf{u} + \mathbf{b} \right) - \gamma \mathbf{x} - \alpha \dot{\mathbf{x}} \qquad (4.1)$$

where $\frac{\partial \mathbf{x}}{\partial t} = \dot{\mathbf{x}}$, $\frac{\partial^2 \mathbf{x}}{\partial t^2} = \ddot{\mathbf{x}}$ are the first and second derivatives of the hidden state with respect to time, and $\mathbf{u}$ denotes the input at each time step. The terms $\mathbf{W}_x \mathbf{x}$, $\mathbf{W}_{\dot{x}} \dot{\mathbf{x}}$, and $\mathbf{V} \mathbf{u}$ can then be

4.2 Neural Wave Machines

interpreted as the coupling, damping, and driving terms respectively. Finally, σ is a nonlinear activation function such as the hyperbolic tangent, and γ & α are scalar variables which can be fixed or learned in combination with the above matrices. In practice, the above differential equation can be discretized and integrated numerically using an IMEX (implicit-explicit) discretization scheme shown to preserve the desirable bounds of the continuous system. Such a discretization can be achieved by first introducing a 'velocity' variable $\mathbf{v} = \dot{\mathbf{x}}$, turning the second order system into a set of two coupled first order equations:

$$\dot{\mathbf{x}} = \mathbf{v}, \quad \dot{\mathbf{v}} = \sigma\left(\mathbf{W}_x \mathbf{x} + \mathbf{W}_{\dot{x}} \mathbf{v} + \mathbf{V}\mathbf{u} + \mathbf{b}\right) - \gamma \mathbf{x} - \alpha \mathbf{v} \qquad (4.2)$$

Then, for a fixed time step $0 < \Delta t < 1$, the hidden state $\mathbf{x}$ and velocity $\mathbf{v}$ of the RNN at time $t + 1$ can be updated as:

$$\mathbf{x}_{t+1} = \mathbf{x}_t + \Delta t (\mathbf{v}_{t+1}) \quad \mathbf{v}_{t+1} = \mathbf{v}_t + \Delta t (\mathbf{v}'_t) \qquad (4.3)$$

$$\mathbf{v}'_t = \sigma\left(\mathbf{W}_x \mathbf{x}_t + \mathbf{W}_{\dot{x}} \mathbf{v}_t + \mathbf{V}\mathbf{u}_{t+1} + \mathbf{b}\right) - \gamma \mathbf{x}_t - \alpha \mathbf{v}_t \qquad (4.4)$$

This model was theoretically demonstrated to have a bounded gradient and hidden state magnitude under assumptions on the time-step Δt and the infinity norm of the coupling parameters. Empirically, such stable gradient dynamics were shown to yield better performance than existing RNNs on tasks with very long time-dependencies.

In relation to our goals, the oscillatory dynamics of the coRNN make it amenable to synchronous activity, unlike most existing deep neural network models, and the stable gradient dynamics make it a powerful and flexibly parameterizable sequence model, unlike existing models of traveling waves based on spiking neural networks. However, given that the hidden state $\mathbf{x}$ is not endowed with any notion of spatial layout, it is still not meaningful to study spatiotemporal dynamics in such a model. In the following subsection we describe how such a spatial layout may be implemented efficiently by replacing the fully connected recurrent coupling matrices $\mathbf{W}_x$ and $\mathbf{W}_{\dot{x}}$ with convolution operations.

4.2.2 Local Connectivity

In [17], the authors study a large scale spiking neural network model, quantifying the emergence of traveling waves, and comparing them with waves observed in the human cortex. At a high level, as it is relevant to this work, the study concludes that locally restricted connectivity and distance dependant conduction delays are both necessary and sufficient to produce traveling waves. Further they observe that such waves are fairly robust to the synaptic strengths of their model when given a sufficiently large number of neurons. Given these findings, we hypothesize that the Coupled Oscillitory Recurrent Neural Network may yield traveling waves if similarly constrained.

To impose such constraints we begin by defining an arbitrary topographic layout for the N-dimensional hidden state $\mathbf{x}$ in the model. For computational simplicity, we propose to use a regular 1 or 2 dimensional grid, $\mathbf{x}_{1D} \in \mathbb{R}^{C_h \times N}$ or $\mathbf{x}_{2D} \in \mathbb{R}^{C_h \times \sqrt{N} \times \sqrt{N}}$ respectively, where C_h is the number of simultaneous 'channels' in our hidden state. We then see that specifically, if the recurrent connections $\mathbf{W}_x$ and $\mathbf{W}_{\dot{x}}$ are made local over our spatial dimensions rather than global, and a distance-dependant time-delay introduced, the aforementioned constraints will be satisfied and the remainder of the properties such as synaptic strength and the precise local distribution of connections will be left up to the model to learn. In practice, we simplify the model by restricting the topographic connectivity of each neuron to its immediately adjacent neighbors in the grid, and define all distances (and thus time-delays) to these neurons to be equal to 1. Such a simplification allows us to efficiently implement the local time-delayed connections with a simple size 3 or 3×3 convolutional kernel for 1 and 2 dimensional grids respectively. In summary, our model is then given identically as in Eqs. 4.3 and 4.4 but with convolutional layers in place of the dense recurrent matrices. Explicitly, in the 2-dimensional setting, for convolutional kernels $\mathbf{w}_x, \mathbf{w}_{\dot{x}} \in \mathbb{R}^{C_h \times C_h \times 3 \times 3}$, we get:

$$\mathbf{v}'_t = \sigma \left(\mathbf{w}_x \star \mathbf{x}_t + \mathbf{w}_{\dot{x}} \star \mathbf{v}_t + f_\theta(\mathbf{u}_{t+1}) + \mathbf{b} \right) - \gamma \mathbf{x}_t - \alpha \mathbf{v}_t \tag{4.5}$$

We see we have additionally replaced the linear encoder $\mathbf{V}$ with a function f_θ which can be a convolutional or 'de-convolutional' neural network, or any other mapping from the input to a spatially organized driving force. Importantly, we see that our imposed local connectivity does not immediately invalidate any of the assumptions required for the theorems of Rusch and Mishra [41] about mitigating the EVGP since the infinity norm of the weights is unlikely to significantly increase when simply switching from fully to locally-connected matrices. We include the updated bounds and corresponding proofs in Sect. 4.6. In the end, we denote this model the Neural Wave Machine due to its emergent wave-like dynamics, facilitated by both the oscillatory update equations of the coRNN, and the local connectivity constraints of biological models. In the next section we measure these desired spatiotemporal dynamics of the NWM and further study their impact as an inductive bias on computation.

4.3 Experiments

In the following two subsections we provide experiments which demonstrate: first, that our model learns spatiotemporal structure reminiscent of natural observations from neuroscience; and second, that such structure is beneficial to both efficiency and accuracy. We outline our methods briefly below, and more thoroughly in Sect. 4.5.

4.3.1 Methods

All datasets used in this chapter will be considered as unsupervised unless otherwise noted, and thus we will train the model from Sect. 4.2 as an autoregressive model. To do this, we add a learned decoder from the hidden state $\mathbf{x}_t$ back to the input at the next timestep

4.3 Experiments

$\mathbf{u}_{t+1}$, and train the model with a mean-squared error loss. Explicitly, $\hat{\mathbf{u}}_{t+2} = g_\theta(\mathbf{x}_{t+1})$, and $\mathcal{L} = ||\hat{\mathbf{u}}_{t+2} - \mathbf{u}_{t+2}||_2^2$, where g_θ is the decoder which can again be a convolutional neural network, or any network which maps from the spatial hidden state back to the input space. For the simple tasks in Sect. 4.3.3, and the sequence classification tasks of Sect. 4.3.8 we use minimal encoders and decoders corresponding to single linear layers or small MLPs. For the more complex physical forecasting tasks of Sect. 4.3.8 we use the baseline deep convolutional encoders and decoders defined in the benchmark. As a second minor addition which we observe improves performance on long-term trajectory modeling tasks, we introduce an additional encoder network which learns to predict the initial conditions $\mathbf{x}_0$ and $\mathbf{v}_0$ of the network given a partial 'inference' sequence. Explicitly, we can write this as: $\mathbf{x}_0, \mathbf{v}_0 = f_\theta^{IC}(\{\mathbf{u_t}\}_{t=0}^{T_{inf}})$. Such an initial-condition network is common in the Neural-ODE literature [42], and in this setting it is beneficial to initialize the latent dynamics which would otherwise take a significant number of iterations to reach their final magnitude.

4.3.2 Datasets

To investigate how the NWM's representations change when modeling different datasets, we focus on three training sets in this study. Most simply, we first use a dataset of oriented sine functions (depicted in Fig. 4.2b) with a slowly progressing phase over time steps. This dataset is meant to be a very rough approximation to the spontaneously generated retinal waves observed during development [43]. For this dataset, the wavelength and magnitude of the sine waves are fixed, and sequences are generated by randomly sampling an orientation between 0 to π and then sequentially progressing the phase by $\frac{1}{9}\pi$ for each timestep until two periods are complete. As a second dataset, we borrow the rotating MNIST dataset from the equivariance literature [44], consisting of sequences of MNIST digits with each timestep rotated by an additional $\frac{1}{9}\pi$ radians. This dataset serves to allow us to investigate the existence of generalizable spatio-temporal structure in a limited setting. Finally, for more realistic dynamics, we make use of the recent hamiltonian dynamics suite [19]. At a high

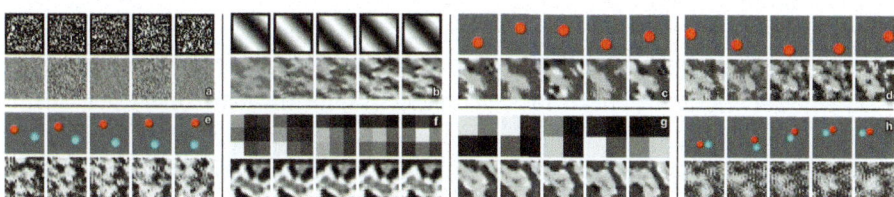

Fig. 4.2 Plot of different datasets used in this work (top) and the associated learned hidden state dynamics (bottom). We see the NWM learns different spatiotemporal structure for each dataset, and no structure when trained on random noise (a). Additional videos of dynamics, and code for experiments, can be found at: github.com/akandykeller/NeuralWaveMachines

level, the benchmark consists of a diversity of tasks governed by known equations of motion, including toy physics examples such as idealized springs, pendulums, orbits, and double-pendulums (Fig. 4.2c, d, e and h), as well as cyclic games (f & g). Models are evaluated based on their ability to accurately forecast dynamics into the future from a limited number of inference frames.

4.3.3 Measuring Spatiotemporal Structure

To measure the spatiotemporal representational structure that the NWM learns, and its alignment with natural structure, we start with the two simplest tasks: modeling simple sine waves, and modeling rotating MNIST digits. We use three separate methods for analyzing the representations learned on these tasks: Cohen's d selectivity metric [45] to depict spatial organization, the Hilbert transform to measure the instantaneous phase and velocity of putative waves [2], and *artificially induced* traveling waves combined with visualized reconstructions to measure the approximate equivalence of latent traveling waves with observed transformations.

4.3.4 Topographic Orientation Selectivity

One of the most common methods to demonstrate spatial organization of neural representations is by measuring their selectivity with respect to different features and plotting this with respect to each neuron's position [46]. As an initial test of a basic form of selectivity, namely orientation selectivity, we consider a hypothesis from the literature about how such structure might arise initially in animals [43]. Specifically, we investigate whether simple periodic inputs, such as the spontaneous retinal waves observed during early development, are sufficient to encourage smooth topographic organization of orientation selectivity when modeled by a minimal NWM. To test this, we train our model on the simple sine waves dataset, and measure the orientation selectivity of each hidden neuron's time-averaged response to a static 36-element sequences of oriented gratings using Cohen's d metric [45]. In Fig. 4.3 we plot the resulting color/angle of maximal d value for each of the 72×72 neurons (or a black x if all $d < 0.65$). We see that the simulated retinal waves do appear to induce topographic organization of orientation selectivity with superficial similarity to the orientation columns of primary visual cortex [47]. Outlined in white, we show a manually identified 'pinwheel' where selectivity for all orientations meet, a hallmark of early visual system organization in many species. In relation to prior models of orientation columns [48], our work does not presuppose the existence of orientation selectivity, but rather it is absent at initialization and it is instead learned in conjunction with topographic organization. We note that the exact statistics of our learned orientation maps have not been measured, and therefore may differ in their current form from those measured in animal studies [49]. In Sect. 4.7.5 we include

4.3 Experiments

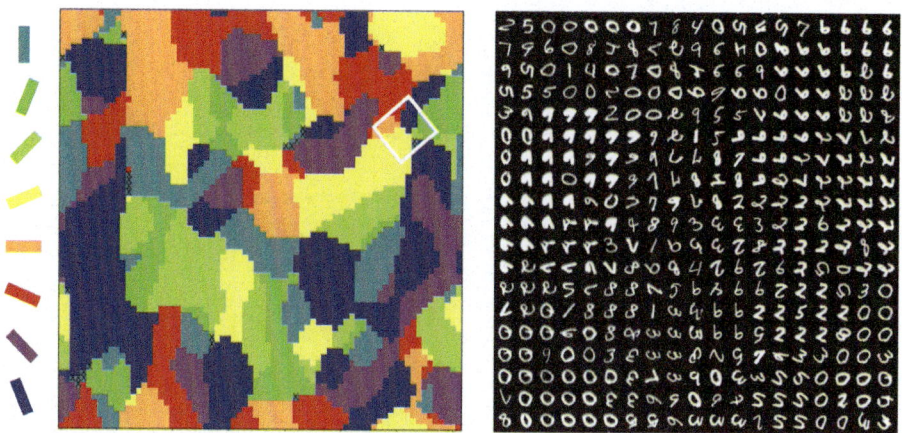

Fig. 4.3 (Left) Plot of orientation selectivity of each NWM hidden neuron **x** after training on simple sine waves. (Right) Plot of the maximum activating image for a subset of NWM hidden neurons after training on the rotating MNIST dataset (See Sect. 4.7.6 for full). We see the NWM learns smooth spatial topographic structure tailored to the input dataset

additional results studying formation mechanism of this orientation selectivity as well as the model parameters which affect the typical length scale of the columns. We leave further precise investigation of the biological similarity to future work.

4.3.5 General Topographic Organization

On the right of Fig. 4.3, we show the spatial structure of feature selectivity for a network trained on rotating MNIST digits instead. Specifically, we plot the image from the MNIST dataset which maximally activates each neuron in our 2-dimensional hidden state (at the final timestep). We see that neurons are organized with respect to digit class and style, but also orientation, implying that activity is likely to travel over these paths as a traveling wave for observed rotation transformations. Such structure is reminiscent of the higher level category selectivity of the ventral temporal cortex [50, 51], and also the temporal structure observed to be related to theta oscillations and waves in the hippocampus [7].

4.3.6 Instantaneous Phase and Velocity

Next, we demonstrate that the proposed model indeed exhibits full spatiotemporal structure beyond static spatial structure. Compared with biological neural networks, it is easy for us to directly visualize the spatio-temporal activity of our network and qualitatively validate the existence of structure. Figures 4.1, 4.2, and 4.4 provide such examples, while additional

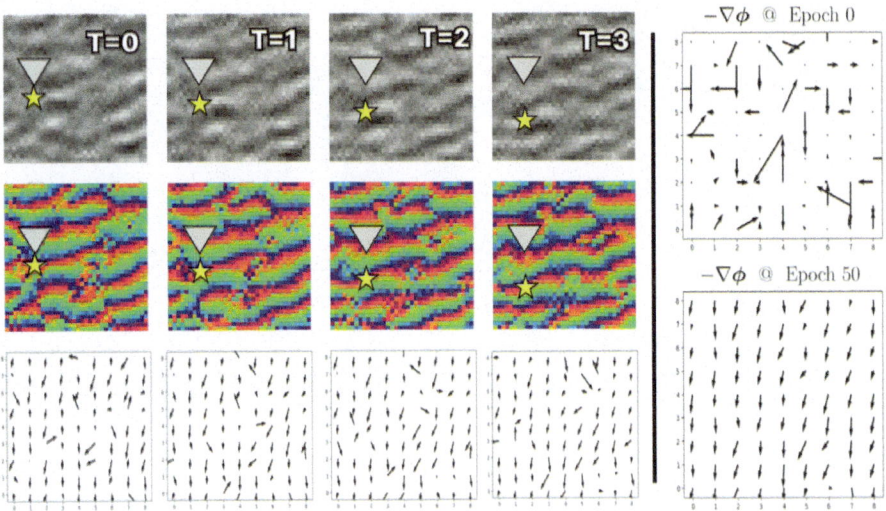

Fig. 4.4 (Left) Plot of hidden state **x** (top), generalized phase ϕ (mid), and estimated wave velocity $-\nabla\phi$ (bot) over the course of a transformation sequence $T = 0$ to 3. A small gold star moves along with a wave front, relative to a stationary grey triangle, both added to help track the approximate peak of a traveling wave in the hidden state. (Right) Estimated wave velocity before and after training

samples can be found in Sect. 4.7.7 and the github repository. For additional rigor, however, we borrow state of the art methods from neuroscience to directly compute the instantaneous phase and velocity of putative waves from noisy real-valued signals. Specifically, we follow the work of [2] and compute the 'generalized phase' of a real valued signal $\mathbf{x}(t)$ by first transforming the signal to a complex-valued analytic signal $\mathbf{x}_a(t)$ through the Hilbert transform $\mathcal{H}$ and then taking the complex argument of this signal as the phase $\phi(t)$ at each point in space and time. Formally: $\mathbf{x}_a(t) = \mathbf{x}(t) + i\mathcal{H}[\mathbf{x}(t)]$, and $\phi(t) = Arg[\mathbf{x}_a(t)]$. Finally, wave velocities can then straightforwardly be computed using the spatial gradient of this phase: $v = -\nabla\phi$. In Fig. 4.4 we depict such phases and velocities for the NWM trained on the rotating MNIST task. We see that, in alignment with expectation, the estimated phases have a spatially periodic pattern which oscillates with sequence length, while the estimated velocities similarly align to point in the downward direction after training (but not before training, as outlined by the disjoint velocity vectors in Fig. 4.4 top right).

4.3.7 Controlled Generation with Induced Traveling Waves

One of the benefits of structured representations in generative models is that they allow for controlled generation of new observations by taking advantage of the known latent operator for a desired input transformation. In this section we demonstrate that such controlled gener-

4.3 Experiments

ation is indeed similarly possible by artificially inducing traveling waves in the NWM hidden state, thereby evidencing the spatiotemporal structure of its representations. Given the high degree of flexibility of the potentially emergent wave dynamics of the 2-D system presented in Fig. 4.4, we concede that two restrictions must be placed on the model in order for us to be able to accurately induce waves which match those the model has learned. Explicitly, we first define the latent space to be a set of disjoint 1-dimensional tori such that learned wave propagation will be restricted to a single axis. Secondly, we restrict our topographic coupling to be 1-directional by masking out all weights except for one (non-central weight) in our convolutional kernel which is shared over all tori. In combination, these restrictions ensure that *if* traveling waves are learned by the model, they will likely be able to be approximately modeled by solutions to the 1-dimensional 1-way wave equation: $y(x, t) = f(x - vt)$.

In Fig. 4.5 we depict the results of this experiment. In detail, we train the 1D NWM described above on a dataset of length $T = 18$ sequences of rotating MNIST digits. At test time, we encode a full sequence (left) and take the final hidden state $\mathbf{x}_T$ as the initial state for our system. We then *induce a traveling wave* in the hidden state in the reverse direction of the instantaneous velocity. In practice, since we have limited our system to 1-dimensional tori, this corresponds to sequentially cyclically shifting (or linearly interpolating) activations across the spatial dimension of each circular subspace according to the inverse of our assumed velocity. The result in Fig. 4.5 (right) shows that indeed by inducing such reverse traveling waves we can then decode the original input sequence, and even predict elements before the start of the sequence (highlighted in pink). Such sensible decodings highlight the generalization power of the representational structure learned by the NWM. In this example we propagate waves with assumed velocity $v = 1$ and observe that this is slightly faster than the ground truth transformation, resulting in a return to the start state in 14 steps rather than 18. Additional transformations can be found in Fig. 4.11 of Sect. 4.7.7.

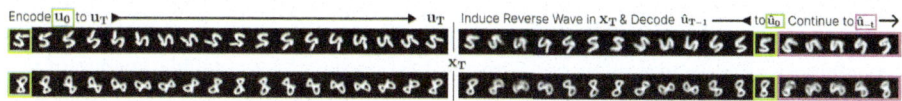

Fig. 4.5 Visualization of controlled generation with induced traveling waves. An input sequence from $\mathbf{u}_0$ to $\mathbf{u}_T$ (left) gets encoded to a hidden state $\mathbf{x}_T$. We then induce a traveling wave in the opposite direction of the estimated instantaneous velocity and observe we can decode back to the original input $\hat{\mathbf{u}}_0$ (highlighted yellow, right). Furthermore, we see by continuing the wave, we can continue the transformation past the bounds of the input sequence (highlighted pink, right)

4.3.8 Computational Implications of Structure

Given the structure measured in Sect. 4.3.3 is known to be related to beneficial inductive biases [11, 44], in this section we perform preliminary experiments to measure such potential benefits in the context of sequence modeling.

4.3.9 An Inductive Bias for Simple Physical Dynamics

First, inspired by the literature relating traveling waves to visual motion perception [8] and spatiotemporal structure in the hippocampus [7], we hypothesize that the spatiotemporal structure of the NWM demonstrated in Sect. 4.3.3 may serve as an inductive bias towards simple physical dynamics. To measure this, we train NWM models on a representative subset of the Hamiltonian dynamics suite, and measure their error when attempting to forecast long test trajectories into the future. Specifically, we consider six distinct dynamic modeling tasks: three simple physical dynamics including the pendulum, spring, and two body gravitational tasks; one less physical but still temporally smooth task, namely the matching pennies task; and the last, the double pendulum, a complex chaotic physical dynamics task. We compare performance of the NWM with the state of the art baselines using optimal hyperparameters directly given in prior work [19, 52]. These include the HGN++ [52], a standard autoregressive model (AR) [53], and a Neural ODE [42] trained both forwards and backwards in time. We additionally include a final globally coupled coRNN baseline with equivalent parameters to our NWM to study the direct impact of the imposed structure on model performance. In Table 4.1 we see that, in alignment with our intuition, the NWM models achieve the lowest forecasting error on the simple physical dynamics tasks, providing evidence in support of the hypothesis that the observed spatiotemporal structure of Sect. 4.3.3 is beneficial for modeling such systems. Further, we see that the coRNN baseline performs the best on the less physical but predictable matching pennies task, while the maximally flexible Neural ODE performs the best on the chaotic double pendulum task. Despite these promising results, we note that accurately measuring forecasting performance in image space is notoriously hard [19, 52], and therefore recommend future work pursue the development of alternative benchmarks and metrics for evaluating the beneficial inductive biases present in the NWM and other forecasting models. In Sect. 4.7.3 and the limitations section below we include additional discussion of these considerations.

4.3.10 Efficiency

As a second potential benefit related to the NWM's demonstrated spatiotemporal structure, our neural wave machines are highly parameter efficient by design when compared to the globally coupled coRNN. As explained in Sect. 4.2, the recurrent connections of our model

4.3 Experiments

Table 4.1 Forward extrapolation mean squared reconstruction error on the Hamiltonian Dynamics Benchmark held-out test set (displayed in units of 1×10^{-8}). We see, in alignment with intuition, the 1 and 2-dimensional Neural Wave Machines (NWM 1D & 2D) perform best on simple physically realistic dynamics such as the spring, pendulum, and two body problem. The globally coupled coRNN performs best on the smooth, but non-physical, matching pennies task, while the maximally flexible Neural ODE performs best on the highly complex and chaotic double pendulum task

	AR	HGN++	ODE	coRNN	NWM 2D	NWM 1D
Spring	20.97	1.58	1.58	2.52	5.46	**1.45**
Pendulum	4,208.0	166.5	166.0	548.0	**110.9**	237.2
Two body	91.4	5.0	4.2	2.0	1.9	**0.9**
Pennies	126.3	190.0	119.3	**28.2**	47.2	43.1
Double pendulum	3,905.0	1,531.0	**1,296.0**	1,666.0	2,512.0	2,821.0

Table 4.2 Test accuracy on supervised sequence benchmarks. All results are mean ± std. over 3 random initalizations

	sMNIST		psMNIST	
	Acc.	#θ (k)	Acc.	#θ (k)
coRNN	99.1 ± 0.1	134	95.0 ± 2.4	134
NWM	98.6 ± 0.3	50	94.8 ± 1.1	50

are restricted to be entirely local as implemented by the convolution operation, thereby allowing for arbitrarily large hidden state sizes with a constant number of recurrent parameters, significantly improving over the quadratically increasing number of parameters in the coRNN. In Table 4.2 we see that on the canonical long sequence classification tasks of sequential MNIST (sMNIST) and permuted sequential MNIST (psMNIST) [41], our model achieves comparable performance with the coRNN (and thus existing state of the art) while requiring a fraction of the parameters. In Sect. 4.7.2 we include additional results on other sequence modeling tasks such as IMDB sentiment classification and long sequence addition showing the same benefits. Interestingly, efficiency in terms of wiring length is also implicated in the formation of orientation columns in natural systems [54]. We believe that our work reinforces this relationship from another perspective by showing that when a recurrent oscillatory computational system is constrained to be wiring length efficient by design, it naturally learns topographic organization (e.g. Fig. 4.3) in order to optimally function.

4.4 Discussion

In this work we introduce the Neural Wave Machine, a recurrent neural network model shown to learn spatiotemporally structured representations through local connectivity and oscillatory dynamics. We propose this model as a rich testing ground for the diversity of computational hypotheses surrounding traveling waves in the neuroscience literature, and demonstrate its potential value in this regard by providing evidence for a variety of hypotheses, including one relating to the origin of orientation columns, and one relating to a simple physical inductive bias. Further, we show that this model is competitive with state of the art on sequence modeling tasks, hoping to encourage future use of such models to study the computational purpose of spatiotemporal dynamics in natural systems.

4.4.1 Related Work

In recent years, multiple works have attempted to integrate topographic organization in deep neural networks for various purposes including learning generalized invariance [55], learning generalized equivariance [44] or for developing more accurate models of the development and structure of natural systems [16, 56, 57]. Other work has studied the temporal aspects of neural activations and attempted to integrate such structure into deep neural networks. For example, researchers have studied the integration of recurrence into feed forward classification networks [58], or the integration spike-time coding through complex activations [59]. Separately, others have aimed to directly integrate natural architectural biases by fixing early layers of a convolutional neural network to mimic the early stages of the natural visual stream, ultimately resulting in improved robustness [60]. Our work is highly related to these efforts in motivation, but largely unique in terms of methodology and its focus on complex spatiotemporal dynamics such as traveling waves. One class of models which shares some relation intuitively is reservoir computing [61]. A primary difference between the NWM and reservoir computing frameworks is that our network has a significant number of learned parameters within its recurrence that mediate complex hidden dynamics, while prior work typically relies on a reservoir of fixed dynamics.

4.4.2 Limitations

In this Chapter we have put significant effort into quantifying the existence of complex spatiotemporal structure and its impact on the NWMs computational performance. However, due to the inherent flexibility of the possible dynamics which may emerge, there remain limitations in our ability to do so. In future work, we would hope to be able to get a more concrete metric corresponding to spatiotemporal structure to better correlate the structure of our models with their performance. Furthermore, on tasks such as forecasting dynamics, it is still an open question how to best compare the performance of such models in the most comprehensive and fair manner [52]. In Sect. 4.7.3 we include additional metrics evaluating model performances on the Hamiltonian Dynamics Suite, highlighting this challenge. Finally, our explorations of parameter efficiency are inherently preliminary and use fully connected encoders and decoders in the NWM, ultimately contributing 45k of the 50k parameters noted for the NWM in Table 4.2. If we were able to replace these components with similarly locally connected functions, such as convolutional networks, the parameter efficiency would further dramatically increase.

4.4.3 Conclusion

As a flexible computational model of traveling waves, we believe the NWM framework offers significant potential to the computational neuroscience community as a method for testing other computational hypotheses relating to traveling waves and synchronous neural dynamics broadly. Similar to convolutional neural networks for modeling the visual system [62–64], neural wave machines do not match all biologically relevant details of neural dynamics, but we believe they may capture sufficient abstract properties to be useful for performing investigations that otherwise wouldn't be possible. Examples of initial hypotheses which we believe would be primarily suited for future study would be the use of traveling waves as a short term memory mechanism [28], or as a mechanism for sequencing actions [31]. Ultimately, we believe this work suggests that complex spatiotemporal dynamics and structure should be investigated further in the future to develop the next set of inductive biases necessary to bring deep neural networks to the same levels of efficiency and robustness that we see in natural intelligence.

4.5 Experiment Details

Videos of traveling waves and code to reproduce all experiments in the Chapter can be found at the following github repository: https://github.com/akandykeller/NeuralWaveMachines.

The code is built as extensions of three existing public repositories, allowing us to reproduce all baseline results from the original authors' code. Specifically, we make use: (I) The coRNN repository (https://github.com/tk-rusch/coRNN) for the supervised sequence experiments, (II) The Topographic VAE repository (https://github.com/akandykeller/TopographicVAE/) for the rotating MNIST experiments, and (III) The DeepMind Physics Inspired Models repository (https://github.com/deepmind/deepmind-research/tree/master/physics_inspired_models) for the Hamiltonian Dynamics Suite Experiments.

4.5.1 Sequence Classification

The efficiency experiments from Sect. 4.3.8 were performed by modifying the published code for the original coRNN [41] to incorporate the local connectivity constraints outlined in the main text. All hyperparameters were thus set to the defaults in the published code which matched the optimal hyperparameters stated by the authors to be found from a grid search on each dataset independently. The baseline coRNN values in Table 4.2 are thus simply from re-running the original authors code, and we observe similar values to those published in [41]. We acknowledge that running a separate grid search for the NWM models may be beneficial to their performance but we were unable to do so due to time and computational constraints and thus leave this to future work. In practice, we found the original coRNN parameters worked well enough to give an initial intuition for the relative performance of the NWM.

For the NWM, the topology of the hidden state was defined to be a regular square 2D grid with side lengths equal to square root of the default hidden state size (or the integer floor of the square root for non-perfect-square values). Each neuron was defined to be connected to its immediate surrounding 8 cells in the grid, in addition to a self-connection. The boundary conditions of the topology were defined to be periodic (implemented through circular padding) such that the global topology was that of a 2-dimensional torus. The recurrent local coupling parameters were shared over all spatial locations of the grid, allowing the above local connectivity to be implemented as a periodic convolution with a kernel of size 3×3. We noted that increasing the number of channels in the convolutional layers dramatically improved performance, and thus for the NWM models in Table 4.2 we use 16 channels in the hidden state. This yielded a parameter count computation of: $\#\theta = 1 \times 256 \times 16 + 16 \times 16 \times 3 \times 3 \times 2 + 256 \times 16 \times 10 = 49,664$.

4.5.2 Rotating MNIST and Sine Waves

The experiments on measuring spatiotemporal structure using the MNIST and simple sine waves datasets were performed by modifying the published code for the Topographic VAE [44] to introduce our proposed NWM in place of the 'shifting temporal coherence' construction of the topographic Student's-T variable in the original paper. To achieve this, the encoder and decoder (f_θ & g_θ) were implemented as a variational autoencoder [65] with a standard Gaussian prior and Bernoulli distribution for the likelihood of the data. Practically, this was achieved by setting the output dimensionality of the encoder f_θ to twice the hidden state dimensionality, defining half of the outputs as the posterior mean μ_θ, and the second half as the log of the posterior variance σ_θ. We additionally found that applying Layer Normalization [66] (denoted LN) to the output of the encoder helped increase convergence speed. Explicitly, the model can thus be described as:

4.5 Experiment Details

$$\mathbf{z}_{t+1} \sim q_\theta(\mathbf{z}_{t+1}|\mathbf{u}_{t+1}) = \mathcal{N}\big(\mathbf{z}_{t+1}; \mu_\theta(\mathbf{u}_{t+1}), \sigma_\theta(\mathbf{u}_{t+1})\mathbf{I}\big), \qquad \bar{\mathbf{z}}_{t+1} = \mathrm{LN}(\mathbf{z}_{t+1}) \quad (4.6)$$

$$\mathbf{v}_{t+1} = \mathbf{v}_t + \Delta t\big(\sigma\left(\mathbf{w}_x \star \mathbf{x}_t + \mathbf{w}_{\dot{x}} \star \mathbf{v}_t + \mathbf{V}\bar{\mathbf{z}}_{t+1} + \mathbf{b}\right) - \gamma \mathbf{x}_t - \alpha \mathbf{v}_t\big) \quad (4.7)$$

$$\mathbf{x}_{t+1} = \mathbf{x}_t + \Delta t\,(\mathbf{v}_{t+1}) \quad (4.8)$$

$$p_\theta(\mathbf{u}_{t+2}|g_\theta(\mathbf{x}_{t+1})) = \mathrm{Bernoilli}(\mathbf{u}_{t+2}; g_\theta(\mathbf{x}_{t+1})) \quad (4.9)$$

where the objective is then computed by averaging the evidence lower bound (ELBO) over the length of the sequence:

$$\mathcal{L}(\mathbf{u}_{1:T}; \boldsymbol{\theta}) = \frac{1}{T}\sum_{t=1}^{T} \mathbb{E}_{\mathbf{z}_t \sim q_\theta(\mathbf{z}_t|\mathbf{u}_t)}\big(\log p_\theta(\mathbf{u}_{t+1}|g_\theta(\mathbf{x}_t)) - D_{KL}[q_\theta(\mathbf{z}_t|\mathbf{u}_t)\|p_\mathbf{Z}(\mathbf{z}_t)]\big) \quad (4.10)$$

The initial conditions for the NWM were then given by simply setting the initial position equal to the first encoder output, and the initial velocity to zero, i.e. $\mathbf{x}_0 = \bar{\mathbf{z}}_0$ & $\mathbf{v}_0 = \mathbf{0}$. Although we did not test the MNIST experiments with a deterministic autoencoder, we note that traveling waves can also clearly be seen in the hidden states of the deterministic models presented in Sects. 4.2 and 4.3.8 (as visualized in Fig. 4.2 and the supplementary material), implying that the variational formulation is not necessary for the emergence of traveling waves.

For the experiment depicted in Fig. 4.4 of Sect. 4.3, we used a simple linear encoder and decoder, and a hidden state dimensionality of 1296 reshaped into a 2D grid of shape 36×36. As in the rest of the Chapter, our topographic connectivity was implemented using a convolutional kernel of shape 3×3 shared over all elements of the grid, with circular padding to enforce periodic boundary conditions on the grid. For training, we presented the model with length 18 sequences of MNIST digits rotating at 20°C per step (thus completing a full period per training sequence). At test time, to create the visualization in Fig. 4.4, we increased the sequence length to 72 elements (or four periods) and visualize a portion of the final period, allowing the system to reach a steady state of wave activity for better visualization. We see that despite not being trained on such long sequences, the NWM is able to generalize and maintain wave activity. For computing the generalized phase, we set use a 4-th order butterworth bandpass filter with bounds set at 0.2 and 0.4 of the Nyquist frequency. As hyperparamters for training, we used standard SGD with momentum of 0.9, a learning rate of 2.5×10^{-4}, and a batch size of 128 for 50 epochs. Following the suggestion outlined in [41], we allowed the parameters γ, α, & Δt to be learned during training by initializing them to $\Delta t = \sigma^{-1}(0.125) = -1.95$, $\gamma = 1.0$, & $\alpha = 0.5$ and then applying appropriate activation functions to keep them within the desired bounds (e.g. sigmoid, ReLU, & ReLU respectively). These hyperparameters and initalization values were determined by implementing a simple toy version of the model with random data and random weights and manually altering parameters to determine the ranges for which coherent wave dynamics were likely to emerge. We note that the properties of the emergent waves appear qualitatively different for different random initalizations of the model. Specifically the wavelength and velocity of the waves appears to vary greatly from run-to-run. We show a few of these different learned dynamics in the additional results section below.

For the experiment depicted in Fig. 4.5 of Sect. 4.3, we used a 3-layer Multi-Layer Perceptron (MLP) for the both encoder and decoder, and a hidden state of dimensionality 1296 reshaped into a set of 24 disjoint 1-D tori (circles) each composed of 54 neurons. We implemented topographic coupling between the immediate neighbors on each circle via a 1-dimensional convolutional kernel of size 3 with circular padding. We then implemented the uni-directionality constraint outlined in the main text be masking the first two elements of the kernel to 0, yielding a kernel with a single trainable parameter explicitly connecting each neuron with its neighbor directly to one side. For training, the dataset and hyperparameters all remained the same as in Fig. 4.4 described above, however the batch size was reduced to 8 for quicker evaluation. We found that additionally adding another layer normalization layer between recurrent steps improved the consistency of the learned waves and thus allowed us to simulate them more accurately at test time. Explicitly this amounted to modifying Eq. 4.8 to: $\mathbf{x}_{t+1} = \text{LN}(\mathbf{x}_t + \Delta t\, (\mathbf{v}_{t+1}))$. Furthermore, to ensure consistency of waves across each circular subspace separately, we shared the bias vector $\mathbf{b}$ across each subspace. To induce a traveling wave in the hidden state of the network and thereby generate the transformation sequence shown in the bottom row of the figure, we first encode the input sequence (shown in the top row), using the equations outlined in this section. We take the final hidden state of the network ($\mathbf{x}_T$) as the initial state from which we begin the wave propagation. Then, across each 1-D circular subspace of the hidden state, we update the values of the hidden state based on the 1-D 1-way wave equation $y(x, t) = f(x - vt)$ for a velocity $v = 1$ for time $t = 1$ to 18. Written in terms of the hidden state $\mathbf{x}_t$, we can effectively propagate waves backwards through the hidden state by moving activation from one spatial location l to a location shifted by $v\Delta t$: $\mathbf{x}_T(l) \to \mathbf{x}_T(l - v\Delta t)$. Practically, this amounts to sequentially circularly shifting the hidden state activation across each circular subspace as depicted in Fig. 4.5.

4.5.3 Hamiltonian Dynamics Suite

The experiments in Sect. 4.3.8 were performed using the DeepMind Physics Inspired Models and Hamiltonian Dynamics Suite, implemented in JAX, as a starting point. All values reported for the baselines (HGN++, AR, and ODE [TR]) were thus obtained by re-running the original code with the hyperparameters stated in [19]. Specifically, for the HGN++, we trained the model both forwards and backwards in time, including over the inference steps, with a final beta value of 0.1 in the ELBO. For the AR model, we used an LSTM with all other parameters default. For the ODE, we used the default parameters with forwards and backwards training, again including inference steps. The only change to the default hyperparamters for all three models was to reduce the batch size to 8 per GPU (thus 32 total per iteration) to fit on our GPUs.

4.6 Analytical Treatment of Neural Wave Machines

The coRNN and NWM architectures were added as extensions to the auto-regressive model already implemented in library. They thus made use of all the same default hyper-parameters, with the only changed values being the aforementioned reduced batch size, an increased number of inference steps (31), an increased number of target steps (60), and an increased hidden state size (23×23). The increased number of inference and target steps was found useful to improve performance on more chaotic tasks such as the pendulum where the accuracy of the initial state is hugely important to the model forecasting performance. Additionally, we note that these values are within the values searched by the grid search of the authors in [19] making their use here for comparison relatively fair. The size of the hidden state was picked as the largest which fit in our GPU memory across all devices. The values of α, γ, and Δt were initialized to the same values as the MNIST experiments described above, and were again allowed to be updated during training simultaneously with the other model parameters. For the 2D NWM, the hidden state topology was again defined to be a 2D torus of size (23×23) implemented through periodic convolution with a 3×3 kernel. The 1D NWM topology was similarly composed of 23 disjoint 1D circles each with 23 neurons, again implemented with periodic convolution with a 1×3 kernel. The coRNN and NWM models additionally used a separate initial condition network to initialize $\mathbf{x}_0$ and $\mathbf{v}_0$. This network was implemented as a GRU with a hidden state of size $2 \times 23 \times 23$ which ran backwards over the inference sequence (length 31) first embedded with the model encoder f_θ. The final hidden state of the model was then split in half and taken to initialize the initial positions and velocities of the coRNN & NWMs.

All models make use of the same deep convolutional encoder with ReLU activations and a similarly deep convolutional spatial broadcast decoder as in the original work. They were similarly all trained for 500,000 iterations to match the original work.

4.5.4 Hardware Details

All models were run on a cluster across roughly 8 NVIDIA GeForce 1080Ti GPUs, 8 NVIDIA GeForce 980Ti GPUs, and 8 NVIDIA Titan X GPUs. Each model in Table 4.1 thus required roughly 6–8 GPU days to train to the final number of iterations.

4.6 Analytical Treatment of Neural Wave Machines

In this section we extend the analytical treatment of Neural Wave Machines, verifying that the model does indeed inherit many of the same beneficial bounds on hidden state and gradient magnitudes as the original coRNN, as stated in the main text. Specifically, by carefully reviewing the proofs for Proposition 3.1 (bounded hidden state energy) and Proposition 3.2 (bounded hidden state gradients) of [41], it can be shown that the Neural Wave Machine satisfies the conditions necessary for these bounds to similarly hold with minor modifications. At a high level, the intuition for why these bounds hold is that our convolutional parameterization of the coupling matrices does not change the theoretical bounds on the infinity norm of the weights, the crucial element necessary for bounding these quantities (e.g. see equation (13) of [41]). In the following, we detail each of these bounds more precisely.

4.6.1 Bounds on Hidden State Energy

Identically following the proof of Proposition 3.1, from Section E.1 of [41], defining the total energy of our model's hidden state as $x_n^T x_n + v_n^T v_n$, it can be seen this value is bounded at time-step n, and with hidden state size m, as:

$$x_n^T x_n + v_n^T v_n \leq x_0^T x_0 + v_0^T v_0 + nm\Delta t$$

We see that this bound does not change from the original work as the derivation is not dependent on the parameterization of the coupling matrices $\mathbf{W}, \mathcal{W}$. Furthermore, this bound applies equally in the case when we have non-zero initial conditions (as through our initial condition network).

4.6.2 Sensitivity to Inputs

From Section E.2, Proposition E.1, of [41], it can be seen that the NWM also inherits a bound on how much differences in inputs are able to change the hidden state. Specifically, since the activation function we use is tanh, our bound is identical. This is the theoretical justification for our comment regarding the NWM's apparent inability to model chaotic dynamics (which we expand on in Sect. 4.7.4).

4.6.3 Bounds on Hidden State Gradient

From Section E.3, following the proof of Proposition 3.2, of [41], we see that, assuming $\alpha = \gamma = 1$, we can again derive bounds on the gradient of the loss with respect to the model parameters. Specifically, the outline of the proof is nearly identical, with only equation (28) being modified to reflect the fact that our parameters are now shared over all spatial locations (due to the convolution). In detail, the matrix $\mathbf{Z}_{m,\tilde{m}}^{i,j}$ no longer only has a single non-zero value, but instead m non-zero values equal to $\sigma'(\mathbf{A}_{k-1})_i$ (for an m sized hidden state). We see that when this matrix is then multiplied with each vector $(\mathbf{x}_{k-1}, \mathbf{v}_{k-1}, \mathbf{u}_k)$, using the bound $||\mathbf{Z}_{m,\tilde{m}}^{i,j}(\mathbf{A}_{k-1})||_\infty \leq 1$, the upper bounds in equation (29) change from $||\mathbf{x}_{k-1}||_\infty, ||\mathbf{v}_{k-1}||_\infty, ||\mathbf{u}_k||_\infty$ to $m||\mathbf{x}_{k-1}||_\infty, m||\mathbf{v}_{k-1}||_\infty, m||\mathbf{u}_k||_\infty$. Carrying these extra factors of m through the rest of the proof, we arrive at the following final bound on the

4.7 Extended Results

gradient of the loss function ξ with respect to any parameter θ:

$$\left|\frac{\partial \xi}{\partial \theta}\right| \leq \frac{3}{2}(m + \bar{X}m^{3/2})$$

where $\bar{X} = \max_n ||\bar{x}_n||_\infty$.

4.6.4 Assumptions

As with the proofs for the coRNN, the same assumptions are necessary for the bounds to hold. Specifically, it is assumed that Δt is chosen such that:

$$\max\left(\frac{\Delta t(1 + ||\mathbf{W}||_\infty)}{1 + \Delta t}, \frac{\Delta t||\mathcal{W}||_\infty}{1 + \Delta t}\right) \leq \Delta t^r, \quad \frac{1}{2} \leq r \leq 1 \quad (4.11)$$

Since this assumption is indeed satisfied throughout training for the original coRNN, we assume that it is likely satisfied with the NWM as well. Intuitively, we find no reason to believe that changing the fully connected matrices $\mathbf{W}$ & $\mathcal{W}$ to convolutional matrices will have the necessary order-of-magnitude impact on the infinity norm of the weight matrices necessary to invalidate this assumption. In preliminary experiments on sMNIST we also find this intuition to hold. Specifically, for the optimal value of $\Delta t = 0.042$, and $r = \frac{1}{2}$, we see that the maximum over training of the quantity of interest (Eq. 4.11) is actually lower for the NWM than the coRNN (0.157 vs. 0.188) with both being lower than the limit (0.205).

4.7 Extended Results

4.7.1 Impact of Δt Parameter

In this section we include an additional preliminary analysis to measure the impact of changing the Δt parameter. In practice, we see that the parameter has an impact not only on the numerical integration, but also on the speed at which the network's hidden state is able to update. Therefore, similar to prior work with the coRNN, we find it best to treat this parameter as a hyperparameter and tune it in addition to the other hypterparameters. In the table below, we show the results of our model on sMNIST for a range of Δt values (Table 4.3):

Table 4.3 Test accuracy on the sMNIST dataset for a range of Δt values

Δt	0.001	0.1	0.042	0.15	0.30	0.45
Test accuracy	87.7	90.6	98.4	97.5	89.8	NaN

Table 4.4 Test accuracy on additional sequence modeling benchmarks including the long-sequence Addition task from [53], and the IMDB sentiment classification task. All results are mean ± std. over 3 random initalizations. We see similar results to those shown in Table 4.2, the NWM achieves comparable performance while requiring significantly fewer parameters

	Adding task		IMDB	
	Accuracy	#θ (k)	Accuracy	#θ (k)
coRNN	0.0035 ± 0.01	131	86.4 ± 0.2	46
NWM	0.0046 ± 0.0016	<1	86.1 ± 0.3	13

We see that a moderate value of Δt is optimal, while too large causes divergence (perhaps due to excessive discretization errors) and too small disrupts information processing in the RNN.

4.7.2 Additional Efficient Sequence Modeling Results

In this section we include additional results comparing the coRNN and NWM on different sequence modeling tasks. Specifically, we show model performance on the long-sequence addition task initially introduced by [53], and the IMDB sentiment classification task [41]. On both datasets we see that the NWM achieves comparable performance to the coRNN while requiring significantly fewer parameters, in line with results on the sMNIIST and psMNIST datasets (Table 4.4).

4.7.3 Additional Hamiltonian Dynamics Results

In this section we include an alternative metric for measuring model forecasting performance on the Hamiltonian Dynamics Suite. Specifically in Table 4.5, we report the 'Valid Prediction Time' as reported in prior work [19], defined as the number of time steps into the future the models are able to accurately predict the dynamics of the system with reconstruction error under a predefined threshold (MSE < 0.025). Given the high variance of the VPT value from batch-to-batch, the values reported in Table 4.5 are computed as the mean and standard deviation of the VPT over the final 5 evaluation iterations. We see that the values roughly agree with those reported in [19], however certain discrepancies may still appear due to the fact that the authors of [19] only report the range of the grid search they performed but not the actual hyperparameter values of their best performing models. Further, we see that the ranking of model performance under this metric is quite noisy due to the high variance of the metric. We therefore urge future work to consider alternative benchmarks and metrics for evaluating the forecasting performance of such models.

4.7 Extended Results

Table 4.5 Valid Prediction Time 'VPT' (± std.) on the Hamiltonian Dynamics Benchmark. We highlight in bold results which fall within one standard deviation of the best performing model. We see that the VPT metric has large standard deviation owing to the reliance on an arbitrary threshold of image-space similarity, however the NWM models still perform favorably compared with existing state of the art

	AR	HGN++	ODE [TR]	coRNN	NWM 2D	NWM 1D
Spring	302 (63)	**447 (0)**	**430 (26)**	375 (14)	311.8 (27)	**431 (24)**
Pendulum	3 (4)	105 (21)	**212 (65)**	179 (91)	155.1 (24)	**174 (65)**
Two body	263 (92)	**444 (3)**	**439 (11)**	431 (40)	413 (53)	420 (27)
Pennies	118 (25)	79 (6)	**164 (14)**	**165 (23)**	141 (37)	**163 (9)**
Double pendulum	0 (0)	11 (5)	**22 (7)**	3 (1)	9 (9)	**10 (8)**

Table 4.6 Test Mean Squared Error of an LSTM and NWM when forecasting the Lorentz '96 attractor. We see that the NWM performs better in the non-chaotic regime ($F = 0.9$), while in chaotic regime ($F = 8$) the LSTM performs significantly better

Model	$F = 0.9$	$F = 8.0$
LSTM	5.2×10^{-3}	1.9×10^{-2}
NWM	2.4×10^{-3}	4.8×10^{-2}

4.7.4 On Modeling Chaotic Dynamics

In this section, we include an extended evaluation to investigate the apparent inability of the NWM to model more chaotic dynamics such as the double pendulum task. To do this, we perform an analogous experiment to that reported in Sect. 4.7.2 of the original coRNN work [41]. Specifically, we measure the ability of our model to predict the state of a system at a fixed 25-time steps ahead for a Lorentz '96 attractor ($x'_j = (x_{i+1} - x_{i-2})x_{i-1} - x_i + F$). Here, F is an external force which controls how chaotic the trajectories are, where $F = 8$ corresponds to a highly caotic trajectory and $F < 1$ is significantly less chaotic. Ultimately, we see that, similar to the original coRNN work, the LSTM performs significantly better than the NWM in the chaotic regime, providing empirical evidence for the theoretical claim that the coupled oscillator networks are unable to model chaotic dynamics (Table 4.6).

4.7.5 On the Formation of Orientation Maps

Although there is significant prior work which can give intuition as to why the smooth orientation selectivity maps of Fig. 4.3 may arise from our model, we believe we are the first to demonstrate a system which actually learns these types of maps from data in the service of sequence modeling. At the highest level, the intuition for the mechanism behind these maps can be seen to come from the combination of phase-synchrony of coupled oscillator systems, and the necessity to model temporally correlated transformations. Extensive prior work on so-called 'phase-reduced' Kuramoto models demonstrates the emergence of complex spatiotemporal patterns such as plane waves, spirals, and pinwheel lattices. Examples include early work from [67] (Fig. 6), showing various steady state phase relationships in the solutions of the locally coupled oscillator dynamics. Similarly, more recent works, [68] (Figs. 3 and 4) and [69] (Figs. 5 and 6) have studied how this phase-locking can vary for different types of chosen couplings. Given that these phase-reduced systems are theoretical approximations to the more flexible (non-reduced) oscillator dynamics implemented in the NWM, it makes sense that we also see these types of phase relationships (e.g. Fig. 4.4 of the main text). When such complex phase-synchrony is combined with the task of sequence modeling, the synchrony can be seen to essentially be inducing local correlations between neurons for each time-step. Thus, when the training set contains input at a variety of different angles, and the model is required to represent these over time, the intuition follows that there will be spatially-smooth orientation selectivity corresponding to these induced correlations. In Fig. 4.6 we provide some quantitative measurements which align with this intu-

 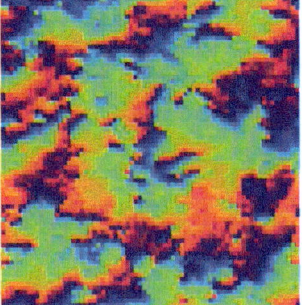

Fig. 4.6 Orientation selectivity (left) and instantaneous phase at a random sequence element (right) for a model trained on the sine waves dataset. We see that the phase synchrony across the neurons is roughly in alignment with the orientation selectivity, supporting the hypothesis that this is one of the primary mechanisms for topographic organization in the NWM

4.7 Extended Results

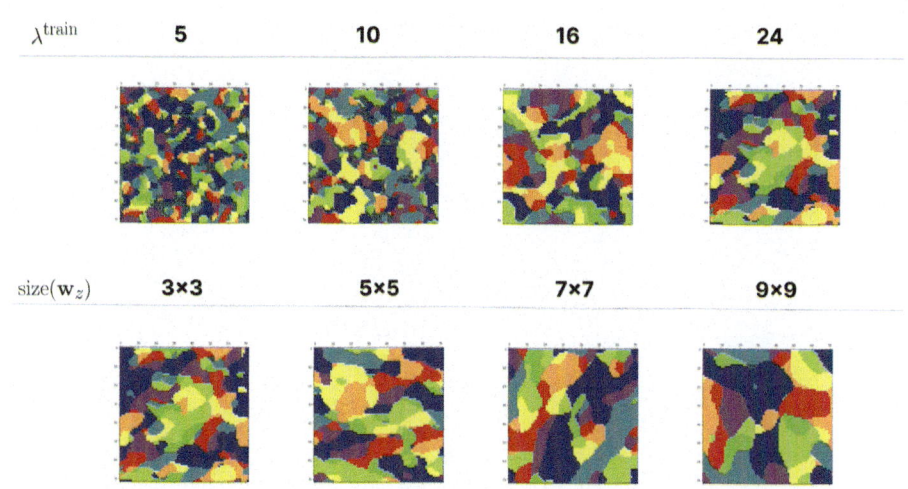

Fig. 4.7 Orientation selectivity maps as a function of training dataset wavelength (λ^{train}), and kernel size (size($\mathbf{w}_z$))

ition. Specifically, the figure shows the instantaneous phase measurement of each neuron (right) next to the orientation selectivity of the same neurons (left). As can be seen, there is a rough correlation between phase values and orientation selectivity, with unexplained variance likely arising due to computing the depicted instantaneous phase values from a single training example, while selectivity measurements are computed over an entire dataset. Furthermore, in Fig. 4.7 we show how different hyperparameters affect the size of the resulting learned orientation columns. We see that both the wavelength of the training dataset (λ^{train} of sine waves) and the kernel size (size($\mathbf{w}_z$)) have a direct increasing relationship with the size of the learned orientation columns, suggesting these parameters could be tuned to better fit observations from neuroscience.

4.7.6 Full Rotating MNIST Topographic Organization

See Fig. 4.8.

Fig. 4.8 Depiction of the maximum activating image for the full set of neurons in the NWM when training on Rotated MNIST. The subset depicted in Fig. 4.3 is highlighted in yellow. We see that topographic organization is widespread and roughly continuous throughout the hidden state

4.7.7 Visualizing Traveling Waves on MNIST

See Figs. 4.9, 4.10 and 4.11.

4.7 Extended Results

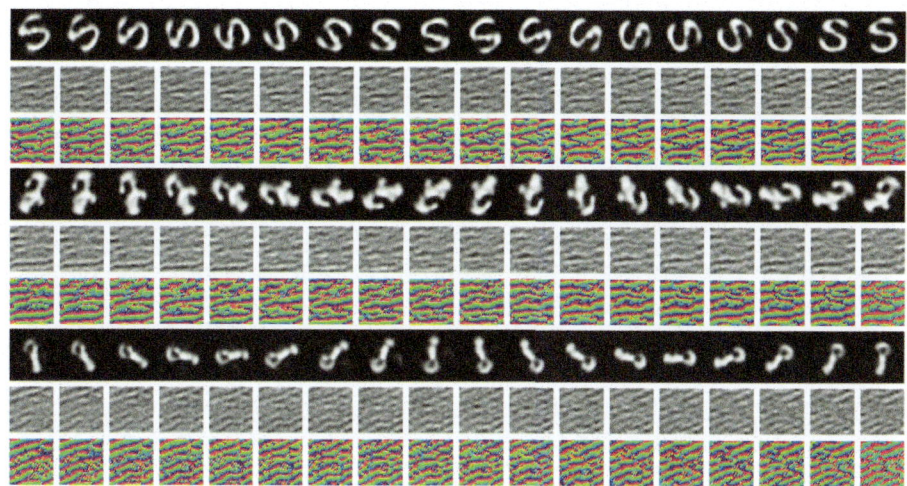

Fig. 4.9 Additional hidden state visualizations for the model in Fig. 4.4. Reconstructions (Top), Hidden state (middle) and generalized phase (bottom), for the final 18 timesteps of the test sequence

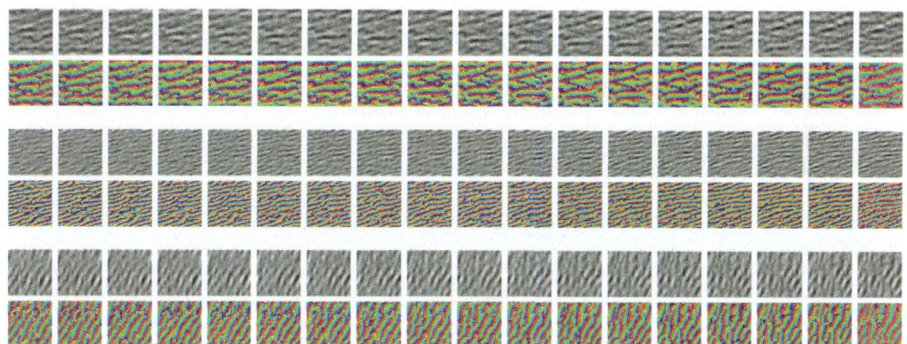

Fig. 4.10 Visualization of the hidden state and phase for three models identical to those in Fig. 4.4, but with different random initalizations. We see that the models learn different wavelengths and velocities depending on their initialization

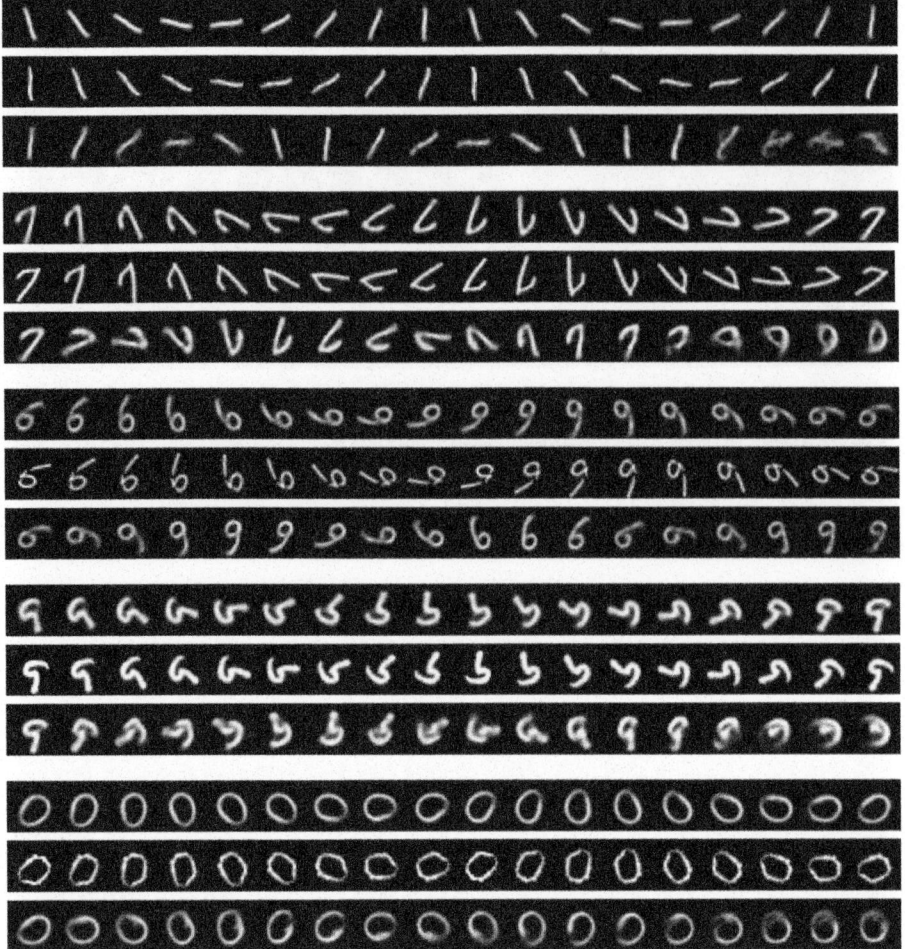

Fig. 4.11 Additional visualizations of reconstructions from induced wave activity in the hidden state of the 1D NWM as depicted in Fig. 4.5. We show a set of random input sequences (top), the original model reconstruction (middle), and images generated by sequentially propagating the initial state backwards by an induced wave and decoding at each step (bottom). We see that, as in the main text, the assumed wave velocity of $v = 1$ is slightly faster than the actual velocity, and thus the reconstructed transformations are slightly faster than the input transformations. Because of this, we also observe that for certain examples, the induced wave reconstructions lose consistency with the input after the first period. This appears to imply that both the initial location of the wave activity matters in addition to its wave properties, and thus our model has learned to only propagate waves over parts of the feature space to optimize the capacity of the hidden state for this dataset. Finally, we observe that the induced transformations occur in reverse order due to the fact that our induced waves propagate in the reverse direction to those naturally exhibited for training examples, effectively propagating backwards in time

References

1. David Wolpert. The lack of a priori distinctions between learning algorithms. *Neural Computation*, 8, 03 1996.
2. Zachary W. Davis, Lyle Muller, Julio Martinez-Trujillo, Terrence Sejnowski, and John H. Reynolds. Spontaneous travelling cortical waves gate perception in behaving primates. *Nature*, 587(7834):432–436, 2020.
3. Lyle Muller, Giovanni Piantoni, Dominik Koller, Sydney S Cash, Eric Halgren, and Terrence J Sejnowski. Rotating waves during human sleep spindles organize global patterns of activity that repeat precisely through the night. *eLife*, 5:e17267, nov 2016.
4. Honghui Zhang, Andrew J. Watrous, Ansh Patel, and Joshua Jacobs. Theta and alpha oscillations are traveling waves in the human neocortex. *Neuron*, 98(6):1269–1281.e4, 2018.
5. Michel Besserve, Nikos Logothetis, and Bernhard Schölkopf. Shifts of gamma phase across primary visual cortical sites reflect dynamic stimulus-modulated information transfer. 01 2015.
6. Lyle Muller, Frédéric Chavane, John Reynolds, and Terrence J. Sejnowski. Cortical travelling waves: mechanisms and computational principles. *Nature Reviews Neuroscience*, 19(5):255–268, 2018.
7. Evgueniy V. Lubenov and Athanassios G. Siapas. Hippocampal theta oscillations are travelling waves. *Nature*, 459(7246):534–539, 2009.
8. Dirk Jancke, Frédéric Chavane, Shmuel Naaman, and Amiram Grinvald. Imaging cortical correlates of illusion in early visual cortex. *Nature*, 2004.
9. Daniel E Worrall, Stephan J Garbin, Daniyar Turmukhambetov, and Gabriel J Brostow. Harmonic networks: Deep translation and rotation equivariance. In *CVPR*, 2017.
10. Taco Cohen and Max Welling. Group equivariant convolutional networks. In *ICML*, 2016.
11. Kunihiko Fukushima. Neocognitron: A self-organizing neural network model for a mechanism of pattern recognition unaffected by shift in position. *Biological Cybernetics*, 36(4):193–202, 1980.
12. Y. Lecun, L. Bottou, Y. Bengio, and P. Haffner. Gradient-based learning applied to document recognition. *Proceedings of the IEEE*, 86(11):2278–2324, 1998.
13. Alex Krizhevsky, Ilya Sutskever, and Geoffrey E Hinton. Imagenet classification with deep convolutional neural networks. In F. Pereira, C.J. Burges, L. Bottou, and K.Q. Weinberger, editors, *Advances in Neural Information Processing Systems*, volume 25. Curran Associates, Inc., 2012.
14. Brenden M. Lake, Tomer D. Ullman, Joshua B. Tenenbaum, and Samuel J. Gershman. Building machines that learn and think like people. *Behavioral and Brain Sciences*, 40:e253, 2017.
15. T. Anderson Keller, Qinghe Gao, and Max Welling. Modeling category-selective cortical regions with topographic variational autoencoders. In *SVRHM 2021 Workshop @ NeurIPS*, 2021.
16. Hyodong Lee, Eshed Margalit, Kamila M. Jozwik, Michael A. Cohen, Nancy Kanwisher, Daniel L. K. Yamins, and James J. DiCarlo. Topographic deep artificial neural networks reproduce the hallmarks of the primate inferior temporal cortex face processing network. *bioRxiv*, 07/2020 2020.
17. Zachary W. Davis, Gabriel B. Benigno, Charlee Fletterman, Theo Desbordes, Christopher Steward, Terrence J. Sejnowski, John H. Reynolds, and Lyle Muller. Spontaneous traveling waves naturally emerge from horizontal fiber time delays and travel through locally asynchronous-irregular states. *Nature Communications*, 12(1), October 2021.
18. Torsten N. Wiesel and David H. Hubel. Ordered arrangement of orientation columns in monkeys lacking visual experience. *Journal of Comparative Neurology*, 158(3):307–318, 1974.
19. Aleksandar Botev, Andrew Jaegle, Peter Wirnsberger, Daniel Hennes, and Irina Higgins. Which priors matter? benchmarking models for learning latent dynamics, 2021.
20. Dr. John R. Hughes. The phenomenon of travelling waves: A review. *Clinical Electroencephalography*, 26(1):1–6, 1995.

21. Tatsuo K. Sato, Ian Nauhaus, and Matteo Carandini. Traveling waves in visual cortex. *Neuron*, 75(2):218–229, 2012.
22. Lyle Muller, Alexandre Reynaud, Frédéric Chavane, and Alain Destexhe. The stimulus-evoked population response in visual cortex of awake monkey is a propagating wave. *Nature Communications*, 5(1), April 2014.
23. Theodoros P Zanos, Patrick J Mineault, Konstantinos T Nasiotis, Daniel Guitton, and Christopher C Pack. A sensorimotor role for traveling waves in primate visual cortex. *Neuron*, 85(3):615–627, 2015.
24. Sayak Bhattacharya, Jacob A Donoghue, Meredith Mahnke, Scott L Brincat, Emery N Brown, and Earl K Miller. Propofol anesthesia alters cortical traveling waves. *Journal of Cognitive Neuroscience*, 34(7):1274–1286, 2022.
25. Karl J. Friston. Waves of prediction. *PLOS Biology*, 17(10):e3000426, 2019.
26. Andrea Alamia and Rufin VanRullen. Alpha oscillations and traveling waves: Signatures of predictive coding? *PLOS Biology*, 17(10):e3000487, 2019.
27. Jean-Rémi King and Valentin Wyart. The human brain encodes a chronicle of visual events at each instant of time through the multiplexing of traveling waves. *The Journal of Neuroscience*, 41(34):7224–7233, 2021.
28. Sayak Bhattacharya, Scott L. Brincat, Mikael Lundqvist, and Earl K. Miller. Traveling waves in the prefrontal cortex during working memory. *PLOS Computational Biology*, 18(1):1–22, 01 2022.
29. Lyle Muller, Giovanni Piantoni, Dominik Koller, Sydney S Cash, Eric Halgren, and Terrence J Sejnowski. Rotating waves during human sleep spindles organize global patterns of activity that repeat precisely through the night. *eLife*, 5, November 2016.
30. Stewart Heitmann and G. Bard Ermentrout. Direction-selective motion discrimination by traveling waves in visual cortex. *PLOS Computational Biology*, 16(9):1–20, 09 2020.
31. Naoyuki Sato. Cortical traveling waves reflect state-dependent hierarchical sequencing of local regions in the human connectome network. *Scientific Reports*, 12(1), January 2022.
32. Anatol Bragin, Gábor Jandó, Zoltán Nádasdy, Jamille Hetke, Kensall Wise, and Gy Buzsáki. Gamma (40-100 hz) oscillation in the hippocampus of the behaving rat. *Journal of neuroscience*, 15(1):47–60, 1995.
33. NE Diamant and A Bortoff. Nature of the intestinal low-wave frequency gradient. *American Journal of Physiology-Legacy Content*, 216(2):301–307, 1969.
34. George Bard Ermentrout and Nancy Kopell. Frequency plateaus in a chain of weakly coupled oscillators, i. *SIAM journal on Mathematical Analysis*, 15(2):215–237, 1984.
35. Yoshiki Kuramoto. Rhythms and turbulence in populations of chemical oscillators. *Physica A: Statistical Mechanics and its Applications*, 106(1-2):128–143, 1981.
36. G Bard Ermentrout and David Kleinfeld. Traveling electrical waves in cortex: insights from phase dynamics and speculation on a computational role. *Neuron*, 29(1):33–44, 2001.
37. Pulin Gong and Cees van Leeuwen. Distributed dynamical computation in neural circuits with propagating coherent activity patterns. *PLOS Computational Biology*, 5(12):1–11, 12 2009.
38. Eugene M Izhikevich and Frank C. Hoppensteadt. Polychronous wavefront computations. *International Journal of Bifurcation and Chaos*, 19(5):1733–1739, 01 2008.
39. Matthew Ricci, Minju Jung, Yuwei Zhang, Mathieu Chalvidal, Aneri Soni, and Thomas Serre. Kuranet: Systems of coupled oscillators that learn to synchronize, 2021.
40. Emre O. Neftci, Hesham Mostafa, and Friedemann Zenke. Surrogate gradient learning in spiking neural networks, 2019.
41. T. Konstantin Rusch and Siddhartha Mishra. Coupled oscillatory recurrent neural network (cornn): An accurate and (gradient) stable architecture for learning long time dependencies. In *International Conference on Learning Representations*, 2021.

42. Ricky T. Q. Chen, Yulia Rubanova, Jesse Bettencourt, and David Duvenaud. Neural ordinary differential equations, 2018.
43. James B. Ackman, Timothy J. Burbridge, and Michael C. Crair. Retinal waves coordinate patterned activity throughout the developing visual system. *Nature*, 490(7419):219–225, 2012.
44. T. Anderson Keller and Max Welling. Topographic vaes learn equivariant capsules, 2021.
45. Jack Cohen. *Statistical Power Analysis for the behavioral sciences*. L. Erlbaum Associates, 1988.
46. David H Hubel and Torsten N Wiesel. Sequence regularity and geometry of orientation columns in the monkey striate cortex. *Journal of Comparative Neurology*, 158(3):267–293, 1974.
47. David H Hubel, Torsten N Wiesel, and Michael P Stryker. Anatomical demonstration of orientation columns in macaque monkey. *Journal of Comparative Neurology*, 177(3):361–379, 1978.
48. N. V. Swindale. A model for the formation of orientation columns. *Proceedings of the Royal Society of London. Series B, Biological Sciences*, 215(1199):211–230, 1982.
49. Matthias Kaschube, Michael Schnabel, Siegrid Löwel, David M. Coppola, Leonard E. White, and Fred Wolf. Universality in the evolution of orientation columns in the visual cortex. *Science*, 330(6007):1113–1116, 2010.
50. Nancy Kanwisher, Josh McDermott, and Marvin M Chun. The fusiform face area: a module in human extrastriate cortex specialized for face perception. *Journal of neuroscience*, 17(11):4302–4311, 1997.
51. Meenakshi Khosla, N. Apurva Ratan Murty, and Nancy Kanwisher. A highly selective response to food in human visual cortex revealed by hypothesis-free voxel decomposition. *Current Biology*, 32(19):4159–4171.e9, 2022.
52. Irina Higgins, Peter Wirnsberger, Andrew Jaegle, and Aleksandar Botev. Symetric: Measuring the quality of learnt hamiltonian dynamics inferred from vision. In M. Ranzato, A. Beygelzimer, Y. Dauphin, P.S. Liang, and J. Wortman Vaughan, editors, *Advances in Neural Information Processing Systems*, volume 34, pages 25591–25605. Curran Associates, Inc., 2021.
53. Jürgen Schmidhuber, Sepp Hochreiter, et al. Long short-term memory. *Neural Comput*, 9(8):1735–1780, 1997.
54. Alexei A Koulakov and Dmitri B Chklovskii. Orientation preference patterns in mammalian visual cortex: a wire length minimization approach. *Neuron*, 29(2):519–527, 2001.
55. Koray Kavukcuoglu, Marc'Aurelio Ranzato, Rob Fergus, and Yann LeCun. Learning invariant features through topographic filter maps. In *2009 IEEE Conference on Computer Vision and Pattern Recognition*, pages 1605–1612, 2009.
56. Fenil R. Doshi and Talia Konkle. Visual object topographic motifs emerge from self-organization of a unified representational space. *bioRxiv*, 2022.
57. Nicholas M. Blauch, Marlene Behrmann, and David C. Plaut. A connectivity-constrained computational account of topographic organization in primate high-level visual cortex. *Proceedings of the National Academy of Sciences*, 119(3):e2112566119, 2022.
58. Tim C. Kietzmann, Courtney J. Spoerer, Lynn K. A. Sörensen, Radoslaw M. Cichy, Olaf Hauk, and Nikolaus Kriegeskorte. Recurrence is required to capture the representational dynamics of the human visual system. *Proceedings of the National Academy of Sciences*, 116(43):21854–21863, 2019.
59. Sindy Löwe, Phillip Lippe, Maja Rudolph, and Max Welling. Complex-valued autoencoders for object discovery, 2022.
60. Joel Dapello, Tiago Marques, Martin Schrimpf, Franziska Geiger, David D. Cox, and James J. DiCarlo. Simulating a primary visual cortex at the front of cnns improves robustness to image perturbations. *bioRxiv*, 2020.
61. Mantas Lukoševičius and Herbert Jaeger. Reservoir computing approaches to recurrent neural network training. *Computer science review*, 3(3):127–149, 2009.

62. Daniel L. K. Yamins, Ha Hong, Charles F. Cadieu, Ethan A. Solomon, Darren Seibert, and James J. DiCarlo. Performance-optimized hierarchical models predict neural responses in higher visual cortex. *Proceedings of the National Academy of Sciences*, 111(23):8619–8624, 2014.
63. Charles F. Cadieu, Ha Hong, Daniel L. K. Yamins, Nicolas Pinto, Diego Ardila, Ethan A. Solomon, Najib J. Majaj, and James J. DiCarlo. Deep neural networks rival the representation of primate it cortex for core visual object recognition. *PLOS Computational Biology*, 10(12):1–18, 12 2014.
64. Nancy Kanwisher, Meenakshi Khosla, and Katharina Dobs. Using artificial neural networks to ask 'why' questions of minds and brains. *Trends in Neurosciences*, 2023.
65. Diederik P Kingma and Max Welling. Auto-encoding variational bayes. *ICLR*, 2014.
66. Jimmy Lei Ba, Jamie Ryan Kiros, and Geoffrey E. Hinton. Layer normalization, 2016.
67. Bard Ermentrout, Alla Borisyuk, Avner Friedman, and David Terman. *Neural Oscillators*, volume 1860, pages 69–106. 01 1970.
68. Seong-Ok Jeong, Tae-Wook Ko, and Hie-Tae Moon. Time-delayed spatial patterns in a two-dimensional array of coupled oscillators. *Phys. Rev. Lett.*, 89:154104, 2002.
69. Michael Breakspear, Stewart Heitmann, and Andreas Daffertshofer. Generative models of cortical oscillations: Neurobiological implications of the kuramoto model. *Frontiers in Human Neuroscience*, 4, 2010.

Part III
Learned Homomorphisms and Disentangled Representations

Latent Traversal as Potential Flows 5

5.1 Motivations

In this section, we outline the diverse set of motivations which provide useful intuition for our method, in addition to outlining clear paths for potential future work.

5.1.1 Traveling Waves in Neuroscience

More abstractly, our work is motivated by the recent interest in traveling waves in the neuroscience literature. Succinctly, traveling waves have recently been observed to exist in a diversity of regions and scales in the biological cortex [1]. Although a consensus has yet to be reached about their exact computational purpose, there is a variety of emerging work which appears to implicate them in the predictive processing of observed transformations from both biological [2–6] and computational perspectives. Specifically, these works suggest that they play the role of integrating information across time, encoding motion, and modulating information transfer. In this work, we leverage these observations to motivate the hypothesis that *traveling waves may be a neural correlate of latent traversals, and thereby serve as an efficient way to encode natural transformations using neural network architectures.* Pursuant to this hypothesis, we expect beneficial performance with physics-inspired PDEs guiding latent traversals in artificial neural networks as well.

5.1.2 Fluid Mechanics as Optimal Transport

Optimal Transport (OT) can be described at a high level as finding a map which moves the probability mass between a source and target distribution with minimal cost. Intuitively, this has a strong connection with latent traversals which can similarly be seen as attempting to move samples from a source probability distribution to a target probability distribution most efficiently while staying on the data manifold. For example, consider aiming to perform a traversal which changes the length of an individual's hair while leaving the rest of their traits unaffected. With the constraint that the traversal must stay on the data manifold, the most efficient traversal would not involve the transformation of multiple variables, as this would require the movement of additional mass, but instead only transform the latent code in a direction which corresponds to the transformation of a single generative factor. In essence, if we were able to learn the underlying structure of the data manifold with respect to various semantic attributes, optimal transport would give us a direct solution to how to perform disentangled traversals.

One method for solving optimal transport problems involves casting them to a fluid mechanical system [7], and solving the associated system numerically. More formally, given the source and target density functions $\rho_0(x), \rho_T(x) \geq 0$, if we construct a dynamical system defined by a continuous density field $\rho(x, t) \geq 0$ and a velocity field $v(x, t)$, where $\rho(x, 0) = \rho_0(x)$ and $\rho(x, T) = \rho_T(x)$, then the classical L_2 Wasserstein distance can be shown to be equal to the infimum of:

$$\sqrt{\int_{\mathbf{R}^d} \int_0^T \rho(x, t) |v(x, t)|^2 \, dx dt} \tag{5.1}$$

over all $v(x, t)$ and $\rho(x, t)$ which satisfy the continuity equation: $\frac{\partial \rho(\mathbf{x},t)}{\partial t} = -\nabla \cdot (v(\mathbf{x}, t) \rho(\mathbf{x}, t))$. For the individual particles which make up this density field, this corresponds to a time-update in the position given by the vector field at their location, i.e.: $\frac{\partial \mathbf{x}}{\partial t} = v(\mathbf{x}, t)$. It turns out that, in terms of the velocity, the optimal solutions to Eq. (5.1) can be written as the gradient of some potential function ϕ, i.e., $v(\mathbf{x}, t) = \nabla_{\mathbf{x}} \phi(\mathbf{x}, t)$, thereby earning the name *potential flows*. Ultimately, by following such a potential flow, the system can be seen to be minimizing the Wasserstein distance, thereby solving the optimal transport problem.

In relation to latent traversals, we see that we can make an intuitive connection between the distribution of points which make up the start and end points of a given semantic traversal (e.g., the distribution of portraits photos with short and long hair respectively), and the source and target distributions in the OT framework. Following such a connection would intuitively suggest that we may be able to learn a corresponding latent potential $\phi(x, t)$ which defines the structure of the latent space with respect to this transformation, and then use the gradient of this field to move particles from one distribution to another.

While making a formal connection with OT remains beyond this Chapter, we see there is still a close intuitive connection between such methods which will be further formalized in the next Chapter. In this Chapter, we mainly present this connection simply as motivation for our method and empirically demonstrate the effectiveness and generality of our approach using this intuition. *One question which comes from this interpretation, is what kind of*

5.2 Learned Potential Flows for Traversal

velocity fields are appropriate for encoding transformations? In the following sections, we provide further intuition that motivates our use of physically-realistic PDEs such as the wave equation to constrain the space-time dynamics of ϕ and the resulting velocity $\nabla \phi$.

5.2 Learned Potential Flows for Traversal

In this section, we present the formulation of our learned potential functions, their integration into generative models under different settings, and the training and sampling strategies. The overview of our method is depicted in Fig. 5.1.

5.2.1 Learning the Potential PDEs

Learning the Potential PDE. Assume we are given a pre-trained generative model $\mathcal{G} : \mathcal{Z} \to \mathcal{X}$ with prior distribution $P_z(z)$. To model K different semantically disentangled latent trajectories, we model each trajectory separately as the gradient of a learned time-dependent scalar potential energy field: $u^k(z_t, t) = \text{MLP}_{\theta^k}([z_t; t]) \in \mathbb{R}$. We use a small multilayer perceptron (MLPs) to learn each potential. The process of traversing from an initial sample (z_0) to a future element (z_t) at time t is then defined as the potential flow $\nabla_z u$ described by this field:

$$z_0 \sim P_z(z) \qquad z_t = z_{t-1} + \nabla_z u^k(z_{t-1}, t-1) \tag{5.2}$$

To encourage the latent potential to model realistic trajectories and follow the intuitions outlined above, we additionally impose a PINN constraint in the form of the second-order wave equation with wave coefficient c:

$$f^k(z_t, t) = \frac{\partial^2}{\partial t^2} u^k(z_t, t) - c^2 \nabla_z^2 u^k(z_t, t) \tag{5.3}$$

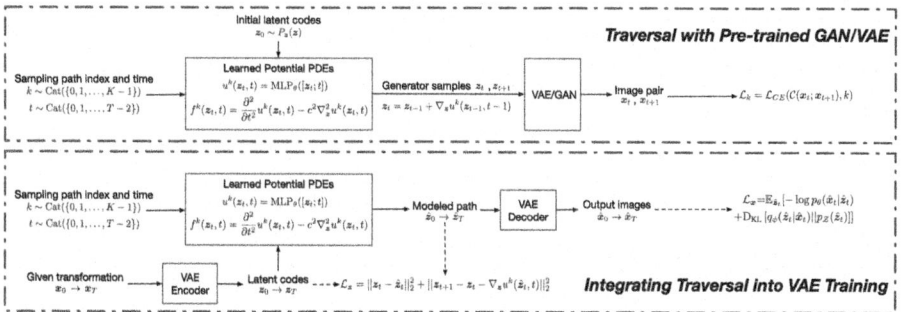

Fig. 5.1 Overview of our learned PDEs for latent traversal in two different experimental settings

Such a constraint makes our potential flow model a good approximation of small amplitude sound waves [8], and empirically is seen to produce highly diverse and realistic trajectories. Our objective is then to minimize:

$$\mathcal{L}_f = \frac{1}{T}\sum_{t=0}^{T-1} \|f^k(z_t, t)\|_2^2, \quad \mathcal{L}_u = \|\nabla_z u^k(z_0, 0)\|_2^2 \tag{5.4}$$

where T represents the total number of timesteps of our latent trajectory, $\mathcal{L}_f$ restricts the energy to obey our physical constraints, and $\mathcal{L}_u$ restricts $u(z_t, t)$ to return no update at $t=0$, thereby matching the initial condition.

Jacobian Regularization. While the above formulation models traversals as physically realistic potential flows, it cannot ensure that the modeled traversal paths are semantically meaningful. Therefore, to make our learned potentials more aligned with the semantics of the data, we take inspiration from prior work and further couple the traversal direction with the Jacobian of the generator. Similar to [9, 10], we first approximate the manipulation on the latent space as

$$\mathcal{G}(z_t + \epsilon \nabla u^k(z_t, t)) \approx \mathcal{G}(z_t) + \epsilon \underline{\frac{\partial \mathcal{G}(z_t)}{\partial z_t} \nabla_z u^k(z_t, t)} \tag{5.5}$$

where ϵ denotes perturbation strength. Intuitively, for sufficiently small ϵ, if the Jacobian-vector product (the underlined term in Eq. (5.5)) can cause large variations in the generated sample, the direction is likely to be semantically meaningful. We therefore introduce a Jacobian-vector product regularization term to encourage the improved semantic variations of our traversals in an unsupervised manner:

$$\mathcal{L}_\mathcal{J} = -\|\frac{\partial \mathcal{G}(z_t)}{\partial z_t} \nabla_z u^k(z_t, t)\|_2^2 \tag{5.6}$$

5.2.2 Integration with Generative Models

5.2.2.1 Traversal with Pre-trained Networks

With pre-trained models, the weights of the generator are frozen. We only update the parameters of our MLPs and of the auxiliary potential-index classifier module. We adopt an auxiliary classifier C to predict the potential index and use the cross-entropy loss to optimize it:

$$\hat{k} = C(x_t; x_{t+1}), \quad \mathcal{L}_k = \mathcal{L}_{CE}(\hat{k}, k) \tag{5.7}$$

where $x_t = \mathcal{G}(z_t)$ is the generated sample from timestep t.

5.2.2.2 Integrating Traversal Into the Training

When training VAEs from scratch, our method can perform "supervised" latent traversal as extra regularization to improve the likelihood. That is, we explicitly model the path of the variations of a semantic attribute during the training process. In this setting, we consider having access to the pre-defined transformation of each variation factor $x_0 \to x_T$. Then we can obtain the corresponding latent codes $z_0 \to z_T$ by feeding images to the encoder, i.e., $z_t = \text{Encode}(x_t)$. Then our potential PDEs manipulate the initial latent codes z_0 to obtain $\hat{z}_1 \to \hat{z}_T$ by progressively performing $\hat{z}_t = z_0 + \sum \nabla_z u^k$. The output images $\hat{x}_1 \to \hat{x}_T$ can be easily attained by decoding $\hat{z}_1 \to \hat{z}_T$. The traversal paths modeled by our wave equations are encouraged to match the ground truth as

$$\begin{aligned} \mathcal{L}_z &= ||z_t - \hat{z}_t||_2^2 + ||(z_{t+1} - z_t) - (\hat{z}_{t+1} - \hat{z}_t)||_2^2 \\ &= ||z_t - \hat{z}_t||_2^2 + ||z_{t+1} - z_t - \nabla_z u^k(\hat{z}_t, t)||_2^2 \end{aligned} \quad (5.8)$$

where the first term penalizes the difference between current latent codes and the ground truth history, and the second term ensures that the future update at the next timestep is realistic. Besides improving the plausibility of traversal paths, we optimize the ELBO:

$$\mathcal{L}_x = \mathrm{e}_{\hat{z}_t}[-\log p_\theta(\hat{x}_t|\hat{z}_t) + \mathrm{D_{KL}}\left[q_\phi(\hat{z}_t|\hat{x}_t)||p_Z(\hat{z}_t)\right]] \quad (5.9)$$

where p_θ parameterizes the generator, and q_ϕ denotes the approximate posterior. The combination of the two losses could yield more structured latent space and more realistic traversal trajectories, which might improve the likelihood.

5.2.2.3 Sampling and Training Strategies

At each training step, we sample a potential with index k from $\text{Cat}(\{0, 1, \ldots, K-1\})$ and a timestep t from $\text{Cat}(\{0, 1, \ldots, T-2\})$. Then we use the selected potential to generate the corresponding velocity fields and obtain the two latent codes z_t and z_{t+1}. Subsequently, the generator is fed with the latent codes and outputs a pair of images x_t and x_{t+1}. Finally, we adopt an auxiliary classifier to predict the potential index $\hat{k}$. The overall loss function is defined as

$$\mathcal{L} = \mathcal{L}_u + \mathcal{L}_f + \underline{\mathcal{L}_\mathcal{J} + \mathcal{L}_k} + \boxed{\mathcal{L}_x + \mathcal{L}_z} \quad (5.10)$$

where $\mathcal{L}_k$ matches the predicted index $\hat{k}$ to the ground truth k, therefore encouraging that each learned potential is significantly distinct and self-consistent to be recognized by a classifier accurately. The boxed terms are only applied to regularize the latent space when integrated into VAE training, while the underlined terms are used for pre-trained models. Notice that different from [11, 12], we do not predict the timesteps from the image pair $[x_t, x_{t+1}]$. This is because our potential PDEs can be very diverse in spatiotemporal form, thus predicting the timesteps from two points on the path demonstrated to be both unnecessary and practically infeasible.

5.3 Experiments

This section starts with the experimental setup, followed by the results under different settings, and ends with in-depth discussions.

5.3.1 Evaluation Methods

Models and Datasets. For experiments of pre-trained GANs, our method is evaluated on SNGAN [13] with AnimeFace [14], BigGAN [15] with ImageNet [16], and StyleGAN2 [17] with FFHQ [18]. For BigGAN, we train the target class "Bernese mountain dog". We adopt LeNet [19] as the auxiliary classifier for SNGAN, while ResNet-18 [20] based classifier is used for both BigGAN and StyleGAN2. For the VAEs experiments, we use the VAE encoder as the auxiliary classifier and evaluate our method on MNIST [21] and dSprites [22] datasets.
MLP for Modeling PDEs. We use sinusoidal positional embeddings [23] to embed the timestep t. Linear layers with Tanh activations are used for embedding the latent code input z. Another linear layer is used to fuse features across space and time. We set the wave coefficient c as a learnable parameter and initialize it with 1.
Metrics. For the quantitative evaluation of traversal with GANs, we use Variational Predictability (VP) [24] score and the correlation coefficient between face attributes and traversal steps using pre-trained attribute estimators. The VP score adopts the few-shot learning setting (e.g., 10% images as the training set) to measure the generalization of a simple neural network in classifying the discovered latent directions from a crafted dataset of random image pairs $[x_0, x_T]$. For attribute correlation, we first use S3FD [25] to extract the face region and then compute the normalized Pearson's correlation between potential indexes and traversal steps using several pre-trained attributes estimators, including ArcFace [26] for face identity, FairFace [27] for face attributes (age, race, and gender), and HopeNet [28] for face poses (yaw, pitch, and roll). The correlation results are averaged across 50 random latent samples. For the quantitative evaluation of VAEs, since our method performs vector-based manipulation, traditional single-dimension-based VAE disentanglement metrics such as Mutual Information Gap (MIP) [29] do not apply here. Some works such as [30, 31] can perform the evaluation of quantitative vector-based manipulation but they require supervision of the ground truth. We thus also evaluate the disentanglement performance using the VP score. The log-likelihood over the entire dataset is measured for the experiment of integrating our method into the VAE training.
Baselines. For pre-trained GANs, we compare our method against two representative baselines, i.e., SeFa [32] and WarpedSpace [12]. SeFa uses eigenvectors of the weight matrix

5.3 Experiments

after latent codes for *linear* perturbation, while WarpedSpace *non-linearly* changes the latent codes using the gradients of RBFs. As for VAEs, there are no popular vector-based traversal methods in the literature so we also use WarpedSpace for comparison. Finally, as another controlled baseline, we train a linear function with other settings aligned with our method.

5.3.2 Results with Pre-trained Networks

SNGAN and BigGAN. Figure 5.2 displays the exemplary latent traversal results and the corresponding trajectories with SNGAN and BigGAN. Since the parameters of the generator are frozen, each method would generate the same image for one latent sample. Our PDEs can generate traversal paths with distinct semantics and precise image attribute control, while the baselines suffer from entangled attributes and the non-target semantics also vary during traversal. Moreover, the paths of WarpedSpace are of very limited non-linearity, which is imperceptible unless the non-linear part of the path is significantly amplified. By contrast, our potential PDEs have more diverse shapes and more flexible non-linearity. Table 5.1 presents the quantitative evaluation results of the VP scores. Our PDEs achieve state-of-the-art performance in terms of classification accuracy in the few-shot learning setting. Specifically, our method outperforms the second-best baseline by 7.04% with SNGAN, by 1.22% with BigGAN, and by 12.23% with StyleGAN2. The consistent performance gain on each dataset indicates that the semantics of our traversal paths are indeed more disentangled

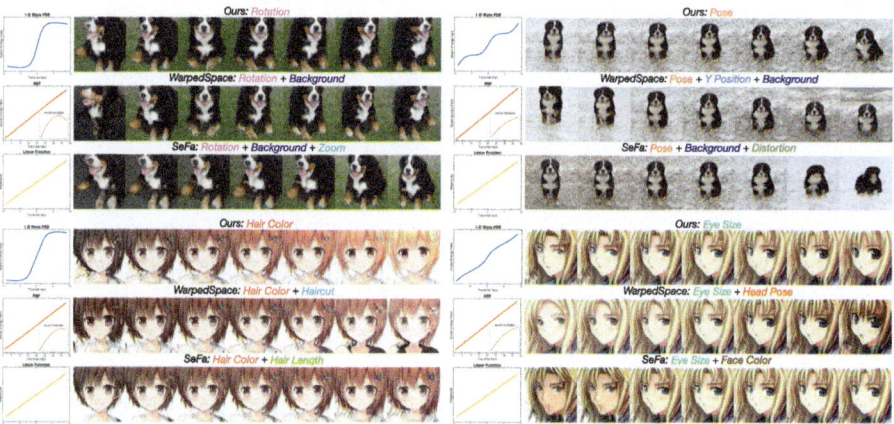

Fig. 5.2 Exemplary traversal paths (potential PDEs for our method) and the corresponding interpolation images with SNGAN and BigGAN. Since the paths of WarpedSpace are of very limited non-linearity that is hard to perceive, we amplify the non-linear part in the sub-figure inside the figure as follows: for a traversal path y of WarpedSpace, we decompose it into $y = y_{LN} + y_{NLN}$ where y_{LN} denotes the linear part and y_{NLN} is the non-linear counterpart. Then the non-linearity part is amplified by $y = y_{LN} + 200 \cdot y_{NLN}$

Table 5.1 Comparison of the VP scores (%) with different GANs averaged over 3 random runs

Models	SeFa	WarpedSpace	Ours
SNGAN	53.76	58.83	**65.89**
BigGAN	13.59	14.07	**15.29**
StyleGAN2	39.20	36.31	**48.54**

than others. It is also worth mentioning that the relatively marginal advantage with BigGAN might stem from the fact that BigGAN generates images in wide domains (1, 000 ImageNet classes). This domain diversity might restrict the actual number of latent semantics, thus limiting the performance.

StyleGAN2. Figure 5.3 compares the exemplary latent traversal with StyleGAN2. The results are coherent with those on SNGAN and BigGAN: the traversal paths of baselines suffer from entangled semantics, while our potential PDEs are able to model trajectories that correspond to more disentangled image attributes. Table 5.2 presents the l_1 normalized correlation results of some common face attributes. As can be seen, most attributes of both SeFa and WarpedSpace have the highest correlation with "identity", implying that their variations of these attributes are often coupled with variations of the face identity during the traversal. By contrast, our method has the best attribute correlations mostly on the diagonal, which explicitly indicates that these attributes of our method are more disentangled from each other.

5.3.3 Results with Pre-trained VAEs

Figure 5.4 displays the exemplary semantics discovered by our method with pre-trained VAEs. Our potential PDEs exhibit a diverse set of different shapes and the interpolation images correspond to distinct transformation factors. Table 5.3 presents the quantitative evaluation of VP scores. The linear baseline and WarpedSpace achieve similar performance, falling behind our method by 4%. This demonstrates again the effectiveness of our PDEs in modelling latent traversal.

Table 5.4 compares the log-likelihood of VAEs integrated with our method. Notice that common disentanglement methods would often sacrifice the likelihood [33]. However, integrating our PDEs into the training process slightly improves the likelihood estimation. Figure 5.5 displays the exemplary traversal results of the pre-defined transformations. Our method is also able to learn and generalize the pre-defined transformation factors well.

5.3 Experiments

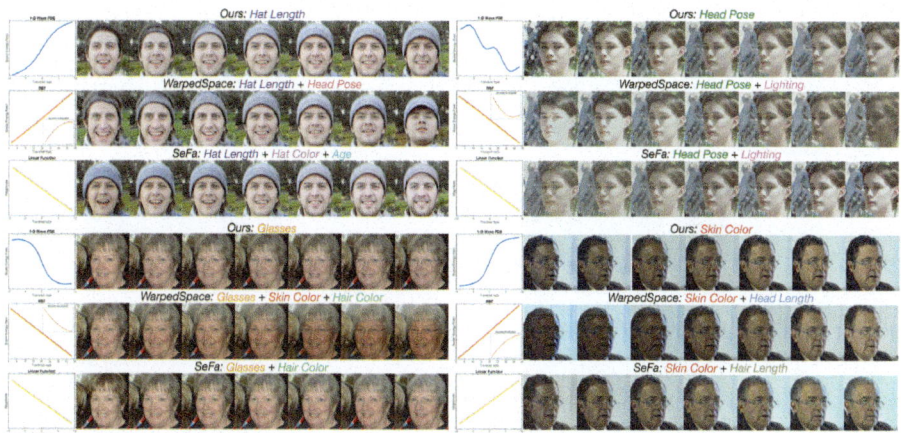

Fig. 5.3 Traversal trajectories (potential PDEs for our method) and the associated interpolation images of the exemplary four attributes with StyleGAN2. The non-linearity of WarpedSpace paths is amplified in the same way as done in SNGAN and BigGAN

Table 5.2 The l_1 normalized attribute correlations of our method (*top*), WarpedSpace (*middle*), and SeFa (*bottom*) based on 50 samples. The second highest correlation is also highlighted if the best value in the row is not on the diagonal

	Yaw	Pitch	Roll	Identity	Age	Race	Gender		Yaw	Pitch	Roll	Identity	Age	Race	Gender		Yaw	Pitch	Roll	Identity	Age	Race	Gender
Yaw	**0.34**	0.09	0.22	0.09	0.03	0.18	0.03	Yaw	**0.34**	0.03	0.05	0.42	0.01	0.08	0.07	Yaw	**0.29**	0.01	0.05	**0.40**	0.04	0.09	0.11
Pitch	0.04	**0.25**	0.11	0.08	0.00	0.08	**0.45**	Pitch	0.01	**0.38**	0.07	0.42	0.01	0.09	0.01	Pitch	0.09	**0.29**	0.06	**0.41**	0.05	0.08	0.01
Roll	0.23	0.19	**0.35**	0.00	0.02	0.03	0.18	Roll	0.10	0.15	0.17	**0.27**	0.02	0.07	**0.22**	Roll	0.03	0.10	0.09	**0.60**	0.00	0.06	**0.12**
Identity	0.01	0.06	0.00	**0.61**	0.21	0.03	0.07	Identity	0.01	0.10	0.00	**0.69**	0.10	0.07	0.01	Identity	0.02	0.05	0.02	**0.74**	0.08	0.08	0.01
Age	0.00	0.06	0.00	0.03	**0.87**	0.00	0.04	Age	0.02	0.09	0.05	**0.52**	**0.25**	0.02	0.05	Age	0.02	0.08	0.02	**0.47**	**0.25**	0.02	0.15
Race	0.05	0.07	0.06	0.02	0.01	**0.73**	0.06	Race	0.05	0.02	0.07	0.12	0.07	**0.54**	0.12	Race	0.07	**0.25**	0.02	**0.58**	0.00	0.00	0.07
Gender	0.08	0.19	0.09	0.04	0.00	0.03	**0.58**	Gender	0.09	0.00	0.02	0.40	0.00	0.00	**0.49**	Gender	0.02	0.05	0.02	**0.43**	0.02	**0.35**	0.12

5.3.4 Results with Networks Trained from Scratch

One interesting geometric property induced by our potential flows is the approximate equivariance for VAEs trained from scratch. At a high level, an equivariant map is one which commutes with a desired transformation group, i.e., $T'[f(x)] = f(T[x])$. This can be understood as preserving geometric symmetries of the input space. The gradient of our potential function can be interpreted as the equivariant latent operator T' corresponding to the observed input transformation $T[x]$. As is typical in the equivariance literature, we can measure how close this is to exact equivariance by measuring the equivariance error:

$$\text{Err} = \sum_{t=1}^{T} |x_t - \hat{x}_t| = \sum_{t=1}^{T} |x_t - \text{Decode}(z_0 + \sum^{t} \nabla_z u^k)| \qquad (5.11)$$

We see this is equivalent to measuring the satisfaction of the equivariance relation $T[x] - f^{-1}(T'[f(x)]) = 0$ where f^{-1} is approximated with the decoder. Table 5.5 presents the

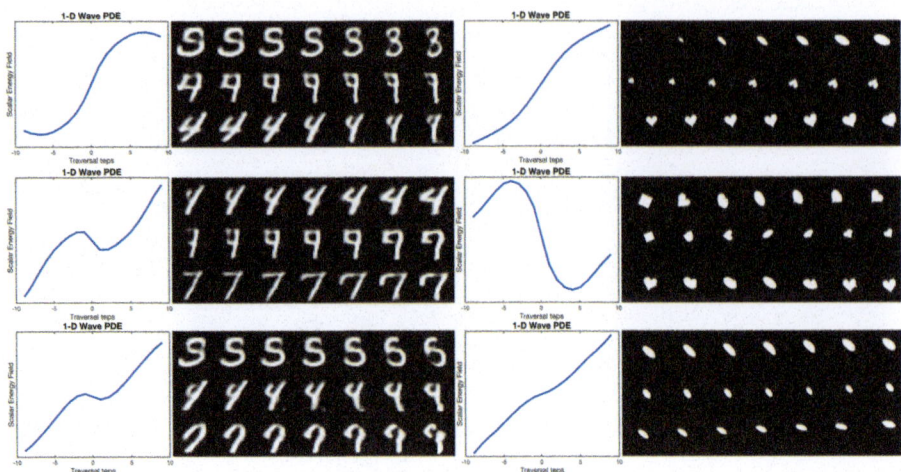

Fig. 5.4 Exemplary semantic attributes and the corresponding traversal trajectories with VAEs trained on MNIST and dSprites

Table 5.3 Comparison of the VP scores (%) with pre-trained VAEs averaged over 3 random runs

Models	WarpedSpace	Ours (Linear)	Ours
MNIST	13.44	12.76	**17.38**
dSprites	15.01	14.25	**18.49**

Table 5.4 The log-likelihood $\log p_\theta(x)$ evaluated over the dataset

Models	Naively trained	Trained with our method
MNIST	−2207.70	**−2144.71**
dSprites	−3848.04	**−3740.97**

evaluation results against a vanilla VAE on transforming MNIST. Note that since the vanilla VAE has no notion of a corresponding transformation in the latent space T' (i.e., no a priori known latent structure), we simply set $\nabla_z u^k$ to 0 and treat this as a lower bound baseline. We see that our method performs significantly above this baseline, indicating that it could be helpful to build equivariant VAEs.

5.4 Discussions

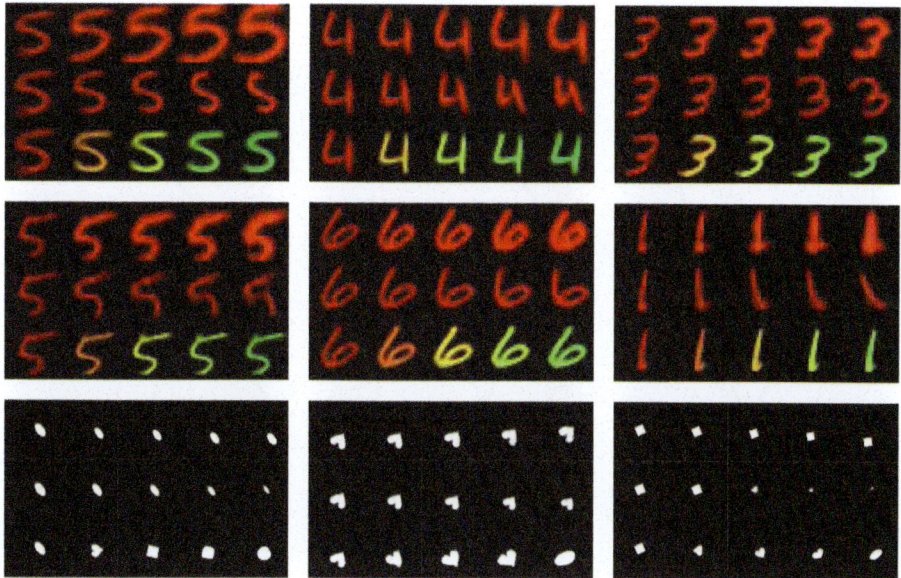

Fig. 5.5 Exemplary traversal results when our method is integrated into the VAE training process. For MNIST, the exhibited transformations are scaling, rotation, and coloring changes from top to bottom. For Despites, the corresponding transformations are y-axis position, scaling, and shape changes from top to bottom

Table 5.5 Equivariance error on MNIST

Transformations	Rotation	Scaling	Coloring
Our method	**235.96**	**230.39**	**240.64**
Vanilla VAE	1278.21	1309.56	1370.54

5.4 Discussions

5.4.1 Flow Path Properties

Linear Directions as Special Cases. We note that the linear traversal approaches can be understood as special cases of our second-order wave equations. Actually, for general linear functions defined as $u(x,t) = a \cdot x + b \cdot t$ where a and b denote the coefficients, the solutions would all correspond to wave equations. In this sense, linear functions are simplified special cases of our waves. One piece of evidence for supporting this is that in certain cases where the structure of the latent space might be simple, our PDEs can also reduce back to functions that are almost linear, such as the traversal paths of the semantic attribute "Eye Size" in Fig. 5.2 and the transformation of scaling in Fig. 5.4 right.

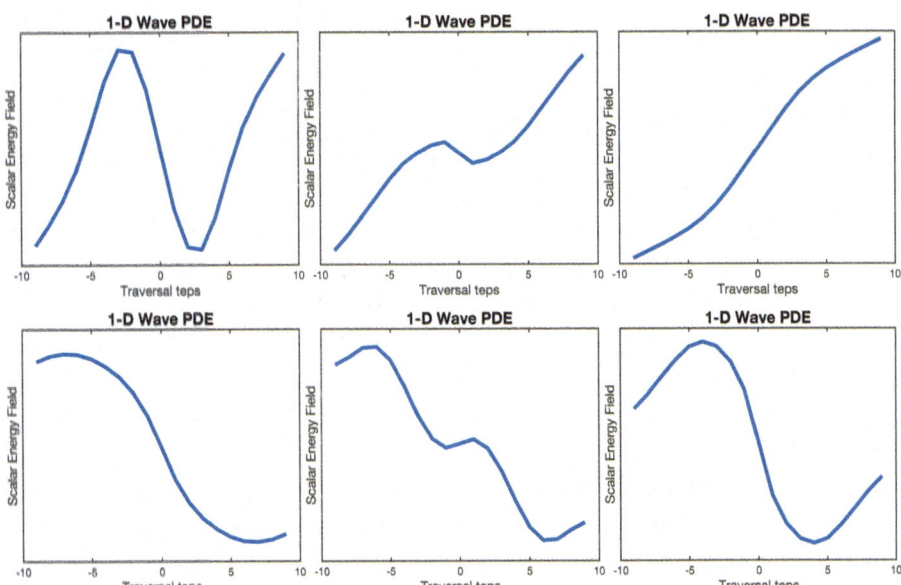

Fig. 5.6 Common shapes of potential PDEs in our experiments

Path Diversity. Our potential PDEs can be very different in shape and period. Figure 5.6 exhibits some common PDEs learned in our experiments. As can be seen, our wave equations allow for a wide set of traversal paths, ranging from linear lines to traveling waves of a full period. This flexibility enables modeling diverse trajectories in the manifold.

Semantic and Trajectory Unambiguity. As shown in Fig. 5.7, for the same traversal path, the semantic attribute is consistent to different samples and the corresponding PDE paths are of very similar shapes. Take the semantic attribute of "Zoom IN" as an example. The scalar potential energy fields of the three images all have slow changes near the endpoints while taking sharp increases in the middle regime. Accordingly, the interpolation images coincide with identical semantics.

Geometric Properties of Latent Spaces. Besides the equivariance property of the encoder/decoder, we also have some novel observations about the shape and variations of $\nabla_z u^k$. For VAEs, we observe that the *simple* variation factors that involve *linear* transformations (e.g., scaling and translation shown in Fig. 5.4 right) tend to be accordingly *more linear* in the latent space. For GANs, the semantic attributes that edit *local* image regions tend to be *more linear* in the latent space, such as the attribute "Eye Size" in Fig. 5.2 and the attributes "Glasses" and "Hat Length" in Fig. 5.3. Through all the experiments, the traversal directions generally tend to have fewer variations when closer to the endpoints. We think this is because at the endpoints (i.e., large timesteps) our potentials learned to not violate the semantic attribute and not to go out of the data manifold.

5.4 Discussions

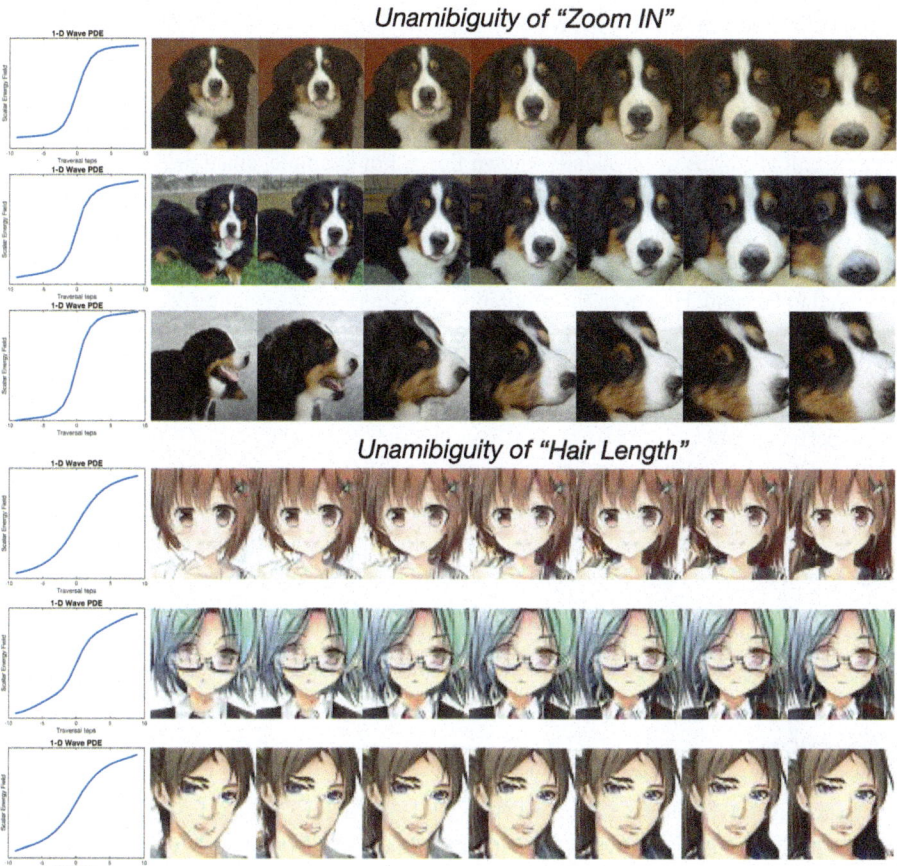

Fig. 5.7 Unambiguity of our potential PDEs and the corresponding discovered semantics: the shape of trajectory and the image attribute of a traversal path are consistent to different samples

5.4.2 Limitations and Future Extensions

Alternative PDE Modeling Approaches. We mainly explore the PINN-based physical constraints to model our PDEs. Despite the flexibility and efficiency, this approach achieves the *soft* PDE constraints approximately. Other alternative possibilities for PDE modeling include Neural Conservation Laws [34] that impose *hard* divergence-free constraints and accurate neural PDE solvers [35, 36]. Investigating other PDE modeling approaches is an important research direction in future work.

Famous PDEs of the Sample Evolution. Driven by our learned velocity field $\nabla u(z, t)$, the sample evolution of z over space and time could satisfy certain PDEs. In particular, with certain $\nabla u(z, t)$, the evolution of z could possibly become some special well-known PDEs, such as heat equations, Fokker Planck equations, and Porous Medium equations.

The specific types depend on the relation between $\nabla u(z,t)$ and $\rho(z,t)$. For instance, if the velocity field is set as $\nabla u(z,t) = -\nabla \log(\rho(z,t))$, the evolution would become the heat equations. More details about the possible relations are kindly referred to [37].

Limitations of Potential Flows. It is known that potential flows are limited in their ability to represent all forms of physically known flows. For example, since the curl of the gradient is known to be zero, potential flows are inherently irrotational and thus cannot model vorticity. In the case of latent traversals, the literature largely appears to model non-cyclic transformations (such as hair length or skin color), and thus this modeling assumption is observed to be valid. However, this limitation explains why the rotation traversals attempted to be learned by our VAE model perform poorly. Ultimately, we propose this framework as a first step towards modeling latent traversals with more complex, physically informed dynamics, and suggest that in some settings, these physical biases may beneficially match the underlying data.

This Chapter proposes a methodology that models latent traversal as potential flows and validates the efficacy in generative models including GANs and VAEs. The connection between disentanglement and approximate learning-based equivariance is unveiled but lacks a probabilistic treatment and formal definitions. In the next chapter, we will move on to present a probabilistic framework that unifies the concepts of disentanglement and approximate equivariance under sequential VAEs.

References

1. Lyle Muller, Frédéric Chavane, John Reynolds, and Terrence J. Sejnowski. Cortical travelling waves: mechanisms and computational principles. *Nature Reviews Neuroscience*, 19(5):255–268, 2018.
2. Dirk Jancke, Frédéric Chavane, Shmuel Naaman, and Amiram Grinvald. Imaging cortical correlates of illusion in early visual cortex. *Nature*, 2004.
3. Tatsuo K Sato, Ian Nauhaus, and Matteo Carandini. Traveling waves in visual cortex. *Neuron*, 2012.
4. Karl J Friston. Waves of prediction. *PLoS biology*, 2019.
5. Andrea Alamia and Rufin VanRullen. Alpha oscillations and traveling waves: Signatures of predictive coding? *PLoS Biology*, 2019.
6. Michel Besserve, Scott C Lowe, Nikos K Logothetis, Bernhard Schölkopf, and Stefano Panzeri. Shifts of gamma phase across primary visual cortical sites reflect dynamic stimulus-modulated information transfer. *PLoS biology*, 2015.
7. Jean-David Benamou and Yann Brenier. A computational fluid mechanics solution to the monge-kantorovich mass transfer problem. *Numerische Mathematik*, 84(3):375–393, 2000.
8. Horace Lamb. *Cambridge mathematical library: Hydrodynamics*. Cambridge University Press, Cambridge, England, 6 edition, 1993.
9. Jiapeng Zhu, Ruili Feng, Yujun Shen, Deli Zhao, Zheng-Jun Zha, Jingren Zhou, and Qifeng Chen. Low-rank subspaces in gans. *NeurIPS*, 2021.
10. Jiapeng Zhu, Yujun Shen, Yinghao Xu, Deli Zhao, and Qifeng Chen. Region-based semantic factorization in gans. *ICML*, 2022.

References

11. Andrey Voynov and Artem Babenko. Unsupervised discovery of interpretable directions in the gan latent space. In *ICML*, 2020.
12. Christos Tzelepis, Georgios Tzimiropoulos, and Ioannis Patras. WarpedGANSpace: Finding non-linear rbf paths in GAN latent space. In *ICCV*, 2021.
13. Takeru Miyato, Toshiki Kataoka, Masanori Koyama, and Yuichi Yoshida. Spectral normalization for generative adversarial networks. *ICLR*, 2018.
14. Brian Chao. Anime face dataset: a collection of high-quality anime faces., 2019.
15. Andrew Brock, Jeff Donahue, and Karen Simonyan. Large scale gan training for high fidelity natural image synthesis. *ICLR*, 2019.
16. Jia Deng, Wei Dong, Richard Socher, Li-Jia Li, Kai Li, and Li Fei-Fei. Imagenet: A large-scale hierarchical image database. In *CVPR*, 2009.
17. Tero Karras, Samuli Laine, Miika Aittala, Janne Hellsten, Jaakko Lehtinen, and Timo Aila. Analyzing and improving the image quality of stylegan. In *CVPR*, 2020.
18. Tero Karras, Samuli Laine, and Timo Aila. A style-based generator architecture for generative adversarial networks. In *CVPR*, 2019.
19. Yann LeCun, Léon Bottou, Yoshua Bengio, and Patrick Haffner. Gradient-based learning applied to document recognition. *Proceedings of the IEEE*, 1998.
20. Kaiming He, Xiangyu Zhang, Shaoqing Ren, and Jian Sun. Deep residual learning for image recognition. In *CVPR*, 2016.
21. Yann LeCun. The mnist database of handwritten digits. 1998.
22. Loic Matthey, Irina Higgins, Demis Hassabis, and Alexander Lerchner. dsprites: Disentanglement testing sprites dataset, 2017.
23. Ashish Vaswani, Noam Shazeer, Niki Parmar, Jakob Uszkoreit, Llion Jones, Aidan N Gomez, Łukasz Kaiser, and Illia Polosukhin. Attention is all you need. *NeurIPS*, 2017.
24. Xinqi Zhu, Chang Xu, and Dacheng Tao. Learning disentangled representations with latent variation predictability. In *ECCV*, 2020.
25. Shifeng Zhang, Xiangyu Zhu, Zhen Lei, Hailin Shi, Xiaobo Wang, and Stan Z Li. S3fd: Single shot scale-invariant face detector. In *ICCV*, 2017.
26. Jiankang Deng, Jia Guo, Niannan Xue, and Stefanos Zafeiriou. Arcface: Additive angular margin loss for deep face recognition. In *CVPR*, 2019.
27. Kimmo Karkkainen and Jungseock Joo. Fairface: Face attribute dataset for balanced race, gender, and age for bias measurement and mitigation. In *WACV*, 2021.
28. Bardia Doosti, Shujon Naha, Majid Mirbagheri, and David J Crandall. Hope-net: A graph-based model for hand-object pose estimation. In *CVPR*, 2020.
29. Ricky TQ Chen, Xuechen Li, Roger B Grosse, and David K Duvenaud. Isolating sources of disentanglement in variational autoencoders. *NeurIPS*, 2018.
30. Georgios Arvanitidis, Lars Kai Hansen, and Søren Hauberg. Latent space oddity: on the curvature of deep generative models. *ICLR*, 2018.
31. Loek Tonnaer, Luis A Pérez Rey, Vlado Menkovski, Mike Holenderski, and Jacobus W Portegies. Quantifying and learning linear symmetry-based disentanglement. *ar*Xiv preprint arXiv:2011.06070, 2020.
32. Yujun Shen and Bolei Zhou. Closed-form factorization of latent semantics in gans. In *CVPR*, 2021.
33. Irina Higgins, Loic Matthey, Arka Pal, Christopher Burgess, Xavier Glorot, Matthew Botvinick, Shakir Mohamed, and Alexander Lerchner. beta-vae: Learning basic visual concepts with a constrained variational framework. *ICLR*, 2016.
34. Jack Richter-Powell, Yaron Lipman, and Ricky TQ Chen. Neural conservation laws: A divergence-free perspective. *NeurIPS*, 2022.
35. Jun-Ting Hsieh, Shengjia Zhao, Stephan Eismann, Lucia Mirabella, and Stefano Ermon. Learning neural pde solvers with convergence guarantees. *ICLR*, 2019.

36. Johannes Brandstetter, Daniel Worrall, and Max Welling. Message passing neural pde solvers. *ICLR*, 2022.
37. Filippo Santambrogio. {Euclidean, metric, and Wasserstein} gradient flows: an overview. *Bulletin of Mathematical Sciences*, 7(1):87–154, 2017.

Flow Factorized Representation Learning 6

6.1 Factorized Representation Learning

We provide an alternative viewpoint at the intersection of these two fields of work which we call Flow Factorized Representation Learning. Figure 6.1 depicts the high-level illustration. Given k different transformations $p_k(x_t|x_0)$ in the input space, we have the corresponding latent probabilistic path $\int_{z_0,z_t} q(z_0|x_0)q_k(z_t|z_0)p(x_t|z_t)$ for each of the transformations. Each latent flow path $q_k(z_t|z_0)$ is generated by the gradient field of some learned potentials ∇u^k following fluid mechanical dynamic Optimal Transport (OT) [1]. Our framework allows for a novel perspective on both *disentanglement* and *equivariance*. Below we detail these proposed definitions.

6.1.1 Learned Equivariance

The concept of equivariance in our framework means that the two probabilistic paths, i.e., $p_k(x_t|x_0)$ in the image space and $\int_{z_0,z_t} q(z_0|x_0)q_k(z_t|z_0)p(x_t|z_t)$ in the latent space, would eventually result in the same distribution of transformed data.

6.1.2 Disentanglement

The definition of disentanglement refers to the distinct set of tangent directions ∇u^k that follow the OT paths to generate latent flows for modeling different factors of variation. The tangent bundle of disentangled directions mean that all the probability paths following our learned flows are orthogonal at every point in the latent space, and the OT property guarantees that the initial distribution efficiently evolves to the terminal via the optimal probabilistic paths.

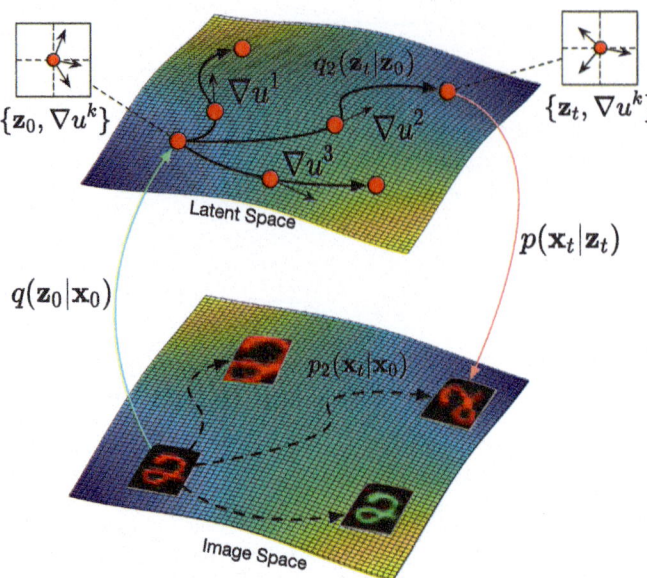

Fig. 6.1 Illustration of our flow factorized representation learning: at each point in the latent space we have a distinct set of tangent directions ∇u^k which define different transformations we would like to model in the image space. For each path, the latent sample evolves to the target on the potential landscape following dynamic optimal transport

6.2 The Generative Model

In this section, we first introduce our generative model of sequences and then describe how we perform inference over the latent variables in the next section.

6.2.1 Flow Factorized Sequence Distributions

The model in this chapter defines a distribution over sequences of observed variables. We further factorize this distribution into k distinct components by assuming that each observed sequence is generated by one of the k separate flows of probability mass in latent space. Since in this work we model discrete sequences of observations $\bar{x} = \{x_0, x_1 \ldots, x_T\}$, we aim to define a joint distribution with a similarly discrete sequence of latent variables $\bar{z} = \{z_0, z_1 \ldots, z_T\}$, and a categorical random variable k describing the sequence type (observed or unobserved). Explicitly, we assert the following factorization of the joint distribution over T timesteps:

6.2 The Generative Model

$$p(\bar{x}, \bar{z}, k) = p(k)p(z_0)p(x_0|z_0)\prod_{t=1}^{T} p(z_t|z_{t-1}, k)p(x_t|z_t). \tag{6.1}$$

Here $p(k)$ is a categorical distribution defining the transformation type, $p(x_t|z_t)$ asserts a mapping from latents to observations with Gaussian noise, and $p(z_0) = \mathcal{N}(0, 1)$. A plate diagram of this model is depicted through the solid lines in Fig. 6.2.

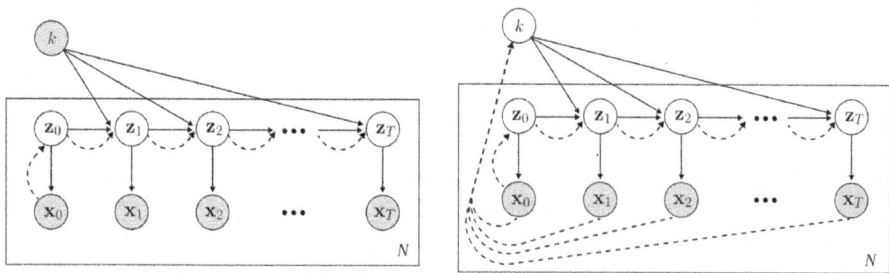

Fig. 6.2 Depiction of our model in plate notation. (Left) Supervised, (Right) Weakly-supervised. White nodes denote latent variables, shaded nodes denote observed variables, solid lines denote the generative model while dashed lines denote the approximate posterior. We see, as in a standard variational autoencoder, our model approximates the initial one-step posterior $p(z_0|x_0)$, but additionally approximates the conditional transition distribution $p(z_t|z_{t-1}, k)$ through optimal transport over an approximate potential landscape

6.2.2 Prior Time Evolution

To enforce that the time dynamics of the sequence define a proper flow of probability density, we compute the conditional update $p(z_t|z_{t-1}, k)$ from the continuous form of the continuity equation: $\partial_t p(z) = -\nabla \cdot (p(z)\nabla\psi^k(z))$, where $\psi^k(z)$ is the kth potential function which advects the density $p(z)$ through the induced velocity field $\nabla\psi^k(z)$. Considering the discrete particle evolution corresponding to this density evolution, $z_t = f(z_{t-1}, k) = z_{t-1} + \nabla_z\psi^k(z_{t-1})$, we see that we can derive the conditional update from the continuous change of variables formula [2, 3]:

$$p(z_t|z_{t-1}, k) = p(z_{t-1})\left|\frac{df(z_{t-1}, k)}{dz_{t-1}}\right|^{-1} \tag{6.2}$$

In this setting, we see that the choice of ψ ultimately determines the prior on the transition probability in our model. As a minimally informative prior for random trajectories, we use a diffusion equation achieved by simply taking $\psi^k = -D_k \log p(z_t)$. Then according to the continuity equation, the prior evolves as:

$$\partial_t p(z_t) = -\nabla \cdot \left(p(z_t) \nabla \psi \right) = D_k \nabla^2 p(z_t) \qquad (6.3)$$

where D_k is a constant coefficient that does not change over time. The density evolution of the prior distribution thus follows a constant diffusion process. We set D_k as a learnable parameter which is distinct for each k.

6.3 Flow Factorized Variational Autoencoders

To perform inference over the unobserved variables in our model, we propose to use a variational approximation to the true posterior, and train the parameters of the model as a VAE. To do this, we parameterize an approximate posterior for $p(z_0|x_0)$, and additionally parameterize a set of K functions $u^k(z)$ to approximate the true latent potentials ψ^*. First, we will describe how we do this in the setting where the categorical random variable k is observed (which we call the supervised setting), then we will describe the model when k is also latent and thus additionally inferred (which we call the weakly supervised setting).

6.3.1 Inference with Observed Transformation Categories

When k is observed, we define our approximate posterior to factorize as follows:

$$q(\bar{z}|\bar{x}, k) = q(z_0|x_0) \prod_{t=1}^{T} q(z_t|z_{t-1}, k) \qquad (6.4)$$

We see that, in effect, our approximate posterior only considers information from element x_0; however, combined with supervision in the form of k, we find this is sufficient for the posterior to be able to accurately model full latent sequences. In the limitations section we discuss how the posterior could be changed to include all elements $\{x_t\}_0^T$ in future work.

Combing Eq. (6.4) with Eq. (6.1), we can derive the following lower bound to model evidence (ELBO):

$$\begin{aligned}
\log p(\bar{x}|k) &= \mathbb{e}_{q_\theta(\bar{z}|\bar{x},k)} \left[\log \frac{p(\bar{x}, \bar{z}|k)}{q(\bar{z}|\bar{x}, k)} \frac{q(\bar{z}|\bar{x}, k)}{p(\bar{z}|\bar{x}, k)} \right] \\
&\geq \mathbb{e}_{q_\theta(\bar{z}|\bar{x},k)} \left[\log \frac{p(\bar{x}|\bar{z}, k) p(\bar{z}|k)}{q(\bar{z}|\bar{x}, k)} \right] \\
&= \mathbb{e}_{q_\theta(\bar{z}|\bar{x},k)} \left[\log p(\bar{x}|\bar{z}, k) \right] + \mathbb{e}_{q_\theta(\bar{z}|\bar{x},k)} \left[\log \frac{p(\bar{z}|k)}{q(\bar{z}|\bar{x}, k)} \right]
\end{aligned} \qquad (6.5)$$

6.3 Flow Factorized Variational Autoencoders

Substituting and simplifying, Eq. (6.5) can be re-written as

$$\log p(\bar{x}|k) \geq \sum_{t=0}^{T} e_{q_\theta(\bar{z}|k)}\left[\log p(x_t|z_t, k)\right] - e_{q_\theta(\bar{z}|k)}\left[D_{KL}\left[q_\theta(z_0|x_0)||p(z_0)\right]\right] \\ - \sum_{t=1}^{T} e_{q_\theta(\bar{z}|k)}\left[D_{KL}\left[q_\theta(z_t|z_{t-1}, k)||p(z_t|z_{t-1}, k)\right]\right] \quad (6.6)$$

We thus see that we have an objective very similar to that of a traditional VAE, except that our posterior and our prior now both have a time evolution defined by the conditional distributions.

6.3.2 Inference with Unknown Transformation Categories

When k is not observed, we can treat it as another latent variable, and simultaneously perform inference over it in addition to the sequential latent $\bar{z}$. To achieve this, we define our approximate posterior and instead factorize it as

$$q(\bar{z}, k|\bar{x}) = q(k|\bar{x})q(z_0|x_0) \prod_{t=1}^{T} q(z_t|z_{t-1}, k) \quad (6.7)$$

Following a similar procedure as in the supervised setting, we derive the new ELBO as

$$\begin{aligned} \log p(\bar{x}) &= e_{q_\theta(\bar{z},k|\bar{x})}\left[\log \frac{p(\bar{x}, \bar{z}, k)}{q(\bar{z}, k|\bar{x})} \frac{q(\bar{z}, k|\bar{x})}{p(\bar{z}, k|\bar{x})}\right] \\ &\geq e_{q_\theta(\bar{z},k|\bar{x})}\left[\log \frac{p(\bar{x}|\bar{z}, k)p(\bar{z}|k)}{q(\bar{z}|\bar{x}, k)} \frac{p(k)}{q(k|\bar{x})}\right] \\ &= e_{q_\theta(\bar{z},k|\bar{x})}\left[\log p(\bar{x}|\bar{z}, k)\right] + e_{q_\theta(\bar{z},k|\bar{x})}\left[\log \frac{p(\bar{z}|k)}{q(\bar{z}|\bar{x}, k)}\right] \\ &\quad + e_{q_\gamma(k|\bar{x})}\left[\log \frac{p(k)}{q(k|\bar{x})}\right] \end{aligned} \quad (6.8)$$

We see that, compared with Eq. (6.5), only one additional KL divergence term $D_{KL}\left[q_\gamma(k|\bar{x})||p(k)\right]$ is added. The prior $p(k)$ is set to follow a categorical distribution, and we apply the Gumbel-SoftMax trick [4] to allow for categorical re-parameterization and sampling of $q_\gamma(k|\bar{x})$.

6.3.3 Posterior Time Evolution

As noted, to approximate the true generative model which has some unknown latent potentials ψ^k, we propose to parameterize a set of potentials as $u^k(z,t) = \text{MLP}([z;t])$ and train them through the ELBOs above. Again, we use the continuity equation to define the time evolution of the posterior, and thus we can derive the conditional time update $q(z_t|z_{t-1},k)$ through the change of variables formula. Given the function of the sample evolution $z_t = g(z_{t-1},k) = z_{t-1} + \nabla_z u^k$, we have:

$$q(z_t|z_{t-1},k) = q(z_{t-1})\left|\frac{dg(z_{t-1},k)}{dz_{t-1}}\right|^{-1} \tag{6.9}$$

Converting the above continuous equation to the discrete setting and taking the logarithm of both sides gives the normalizing-flow-like density evolution of our posterior:

$$\log q(z_t|z_{t-1},k) = \log q(z_{t-1}) - \log|1 + \nabla_z^2 u^k| \tag{6.10}$$

The above relation can be equivalently derived from the continuity equation (i.e., $\partial_t q(z) = -\nabla \cdot (q(z)\nabla u^k)$). Notice that we only assume the initial posterior $q(z_0|x_0)$ follows a Gaussian distribution. For future timesteps, we do not pose any further assumptions and just let the density evolve according to the sample motion.

6.3.4 Optimal Transport for Posterior Flow

As an inductive bias, we would like each latent posterior flow to follow the OT path. To accomplish this, it is known that when the gradient ∇u^k satisfies certain PDEs, the evolution of the probability density can be seen to minimize the L_2 Wasserstein distance between the source distribution and the distribution of the target transformation. Specifically, we have:

Theorem 6.1 (Benamou-Brenier Formula [1]) *For probability measures μ_0 and μ_1, the L_2 Wasserstein distance can be defined as*

$$W_2(\mu_0,\mu_1)^2 = \min_{\rho,v}\left\{\int\int \frac{1}{2}\rho(x,t)|v(x,t)|^2\,dx\,dt\right\} \tag{6.11}$$

where the density ρ and the velocity v satisfy:

$$\frac{d\rho(x,t)}{dt} = -\nabla \cdot (v(x,t)\rho(x,t)), \quad v(x,t) = \nabla u(x,t) \tag{6.12}$$

Proof In the fluid mechanical interpretation, the L_2 Wasserstein distance is re-formulated:

$$W^2 = \inf \int_D \int_0^1 \frac{1}{2}\rho(x,t)v(x,t)^2\,dx\,dt \tag{6.13}$$

6.3 Flow Factorized Variational Autoencoders

where the density satisfies the continuity equation ($\partial_t \rho = -\nabla \cdot (\rho(x,t)v(x,t))$). If we introduce the momentum $m(x,t) = \rho(x,t)v(x,t)$ and two Lagrange multipliers u and λ, the Lagrangian function of the Wasserstein distance would be:

$$L(\rho, m, \phi) = \int_D \int_0^1 \frac{||m||^2}{2\rho} + u(\partial_t \rho + \nabla \cdot m) - \lambda(\rho - s^2) \tag{6.14}$$

where the second term is the equality constraint, and the third term is an equality constraint with a slack variable s. Using integration by parts formula, the above equation can be rewritten as

$$L(\rho, m, \phi) = \int_D \int_0^1 \frac{||m||^2}{2\rho} + \int_D u\rho|_0^1 - \int_D \int_0^1 (\partial_t u \rho + \nabla u \cdot m) - \lambda(\rho - s^2) \tag{6.15}$$

Based on the set of Karush–Kuhn–Tucker (KKT) conditions ($\partial_m L = 0$, $\partial_u L = 0$, $\partial_\rho L = 0$, and $\lambda \geq 0$), we would have:

$$\begin{cases} \partial_m L = \frac{m}{\rho} - \nabla u = v - \nabla u = 0 \\ \partial_u L = \partial_t \rho + \nabla \cdot m = 0 \\ \partial_\rho L = -\frac{||m||^2}{2\rho^2} - \partial_t u - \lambda = -\frac{1}{2}||v||^2 - \partial_t u - \lambda = 0 \end{cases} \tag{6.16}$$

where the first condition indicates that the gradient ∇u acts as the velocity field, and the third condition implies the optimal solution is given by the generalized Hamilton-Jacobi (HJ) equation:

$$\partial_t u + \frac{1}{2}||\nabla u||^2 = -\lambda \leq 0 \tag{6.17}$$

We thus apply the generalized HJ equation as the constraints, and further use an extra negative force for more dynamics in modeling the posterior flow:

$$\frac{\partial}{\partial t} u^k(z,t) + \frac{1}{2}||\nabla_z u^k(z,t)||^2 = f(z,t) \quad \text{subject to} \quad f(z,t) \leq 0 \tag{6.18}$$

Here we use another MLP to parameterize the external force $f(z,t)$ and realize the negativity constraint by setting $f(z,t) = -\text{MLP}([z;t])^2$. Notice that here we take the external force as learnable MLPs simply because we would like to obtain a flexible negativity constraint. The MLP architecture is set the same for both $u(z,t)$ and $f(z,t)$. To achieve the PDE constraint, we impose a Physics-Informed Neural Network (PINN) [5] loss as

$$\mathcal{L}_{HJ} = \frac{1}{T} \sum_{t=1}^T \left(\frac{\partial}{\partial t} u^k(z,t) + \frac{1}{2}||\nabla_z u^k(z,t)||^2 - f(z,t) \right)^2 + ||\nabla u^k(z_0, 0)||^2 \tag{6.19}$$

where the first term restricts the potential to obey the HJ equation, and the second term limits $u(z_t, t)$ to return no update at $t=0$, therefore matching the initial condition.

6.4 Experiments

This section starts with the experimental setup, followed by the main qualitative and quantitative results, then proceeds to discussions about the generalization ability to different composability and unseen data, and ends with the results on complex real-world datasets.

6.4.1 Evaluation Methods

Datasets. We evaluate our method on two widely-used datasets in generative modeling, namely MNIST [6] and Shapes3D [7]. For MNIST [6], we manually construct three simple transformations including Scaling, Rotation, and Coloring. For Shapes3D [7], we use the self-contained four transformations that consist of Floor Hue, Wall Hue, Object Hue, and Scale.

Besides these two common benchmarks, we take a step further to apply our method on Falcol3D and Isaac3D [8], two complex *large-scale* and *real-world* datasets that contain sequences of different transformations. Falcol3D consists of indoor 3D scenes in different lighting conditions and viewpoints, while Isaac3D is a dataset of various robot arm movements in dynamic environments.

Baselines. We mainly compare our method with SlowVAE [9] and Topographic VAE (TVAE) [10]. These two baselines could both achieve approximate equivariance. Specifically, TVAE introduces some learned latent operators, while SlowVAE enforces the Laplacian prior $p(z_t|z_{t-1}) = \prod \alpha\lambda/2\Gamma(1/\alpha) \exp\left(-\lambda|z_{t,i} - z_{t-1,i}|^\alpha\right)$ to sequential pairs. Within the disentanglement literature, our method is compared with the supervised PoFlow [11] which adopts a wave-like potential flow for sample evolution, and the unsupervised β-VAE [12] and FactorVAE [13] which encourage independence between single latent dimensions. Finally, the vanilla VAE is used as a controlled baseline.

Metrics. We use the approximate equivariance error $\mathcal{E}_k$ and the log-likelihood of transformed data $\log p(x_t)$ as the evaluation protocols. The equivariance error is defined as $\mathcal{E}_k = \sum_{t=1}^T |x_t - \text{Decode}(z_t)|$ where $z_t = z_0 + \sum_{t=1}^T \nabla_z u^k$. For TVAE, the latent operator is changed to $\text{Roll}(z_0, t)$. For unsupervised disentanglement baselines [12, 13] and SlowVAE [9], we carefully select the latent dimension and tune the interpolation range to attain the traversal direction and range that correspond to the smallest equivariance error. Since the vanilla VAE does not have the corresponding learned transformation in the latent space, we simply set $\nabla_z u^k = 0$ and take it as a lower-bound baseline. For all the methods, the results are reported based on 5 runs. Notice that the above equivariance error is defined in the output space. Another reasonable evaluation metric is instead measuring error in the latent space as $\mathcal{E}_k = \sum_{t=1}^T |\text{Encode}(x_t) - z_t|$. We see the first evaluation method is more comprehensive as it further involves the decoder in the evaluation.

6.4.2 Learning Equivariant Latent Flows

Qualitative results. Figures 6.3 and 6.4 display decoded images of the latent evolution on MNIST [6] and Shapes3D [7], respectively. On both datasets, our latent flow can perform the target transformation precisely during evolution while leaving other traits of the image unaffected. In particular, for the weakly-supervised setting, the decoded images (i.e., the bottom rows of Figs. 6.3 and 6.4) can still reproduce the given transformations well and it is even hard to visually tell them apart from the generated images under the supervised setting. This demonstrates the effectiveness of the weakly-supervised setting of our method, and implies that qualitatively our latent flow is able to learn the sequence transformations well under both supervised and weakly-supervised settings.

Quantitative results. Tables 6.1 and 6.2 compare the equivariance error and the log-likelihood on MNIST [6] and Shapes3D [7], respectively. Our method learns the latent flows which model the transformations precisely, achieving the best performance across datasets under different supervision settings. Specifically, our method outperforms the previous best baseline by 69.74 on average in the equivariance error and by 32.58 in the log-likelihood on MNIST. The performance gain is also consistent on Shapes3D: our method surpasses the second-best baseline by 291.70 in the average equivariance error and by 120.42 in the log-likelihood. In the weakly-supervised setting, our method also achieves very competitive performance, falling behind that of the supervised setting in the average equivariance error slightly by 6.22 on MNIST and by 67.88 on Shapes3D.

Disentanglement scores. Following the previous chapter, we take the VP scores [15] as the disentanglement metric because they do not pose any assumptions on the latent space but only require image pairs $[x_0, x_T]$ of different transformations for evaluation. The VP

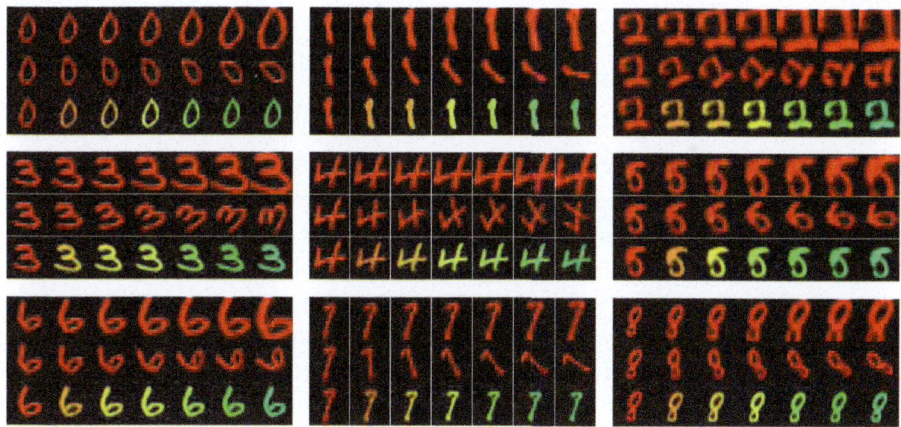

Fig. 6.3 Exemplary latent evolution results of Scaling, Rotation, and Coloring on MNIST [6]. The top two rows are based on the supervised experiment, while the images of the bottom row are taken from the weakly-supervised setting of our experiment

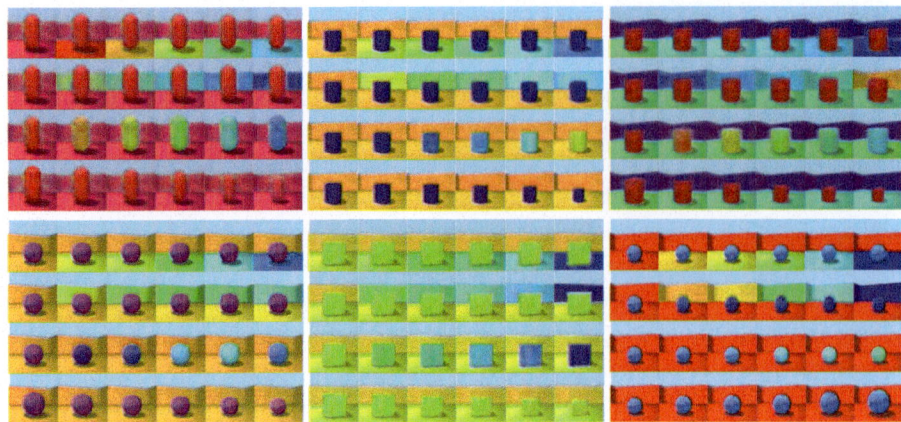

Fig. 6.4 Exemplary latent flow results on Shapes3D [7]. The transformations from top to bottom are Floor Hue, Wall Hue, Object Hue, and Scale, respectively. The images of the top row are from the supervised experiment, while the bottom row is based on the weakly-supervised experiment

Table 6.1 Equivariance error $\mathcal{E}_k$ and log-likelihood $\log p(x_t)$ on MNIST [6]

Methods	Supervision?	Equivariance error ($\downarrow$)			Log-likelihood ($\uparrow$)
		Scaling	Rotation	Coloring	
VAE [14]	No (✗)	1275.31 ± 1.89	1310.72 ± 2.19	1368.92 ± 2.33	−2206.17 ± 1.83
β-VAE [12]	No (✗)	741.58 ± 4.57	751.32 ± 5.22	808.16 ± 5.03	−2224.67 ± 2.35
FactorVAE [13]	No (✗)	659.71 ± 4.89	632.44 ± 5.76	662.18 ± 5.26	−2209.33 ± 2.47
SlowVAE [9]	Weak ( )	461.59 ± 5.37	447.46 ± 5.46	398.12 ± 4.83	−2197.68 ± 2.39
TVAE [10]	Yes (✓)	505.19 ± 2.77	493.28 ± 3.37	451.25 ± 2.76	−2181.13 ± 1.87
PoFlow [11]	Yes (✓)	234.78 ± 2.91	231.42 ± 2.98	240.57 ± 2.58	−2145.03 ± 2.01
Ours	Yes (✓)	**185.42 ± 2.35**	**153.54 ± 3.10**	**158.57 ± 2.95**	**−2112.45 ± 1.57**
Ours	Weak ( )	193.84 ± 2.47	157.16 ± 3.24	165.19 ± 2.78	−2119.94 ± 1.76

metric adopts the few-shot learning setting (using 1% or 10% of the dataset as the training set) and takes a lightweight neural network for learning to classify image pairs $[x_0, x_T]$ of different attributes. The generalization ability (i.e., validation accuracy) can be thus regarded as a reasonable surrogate for the disentanglement ability. Tables 6.3 and 6.4 present the VP scores of all the baseline methods on MNIST and Shapes3D. To ensure a fair comparison, for FactorVAE and β-VAE, we choose the dimensions with the lowest equivariance errors to generate image pairs of different transformations. Our method outperforms the previous disentanglement baselines and achieves superior performance on the VP scores. This indicates that our flow-factorized VAE has better disentanglement ability.

Table 6.2 Equivariance error $\mathcal{E}_k$ and log-likelihood $\log p(\boldsymbol{x}_t)$ on Shapes3D [7]

Methods	Supervision?	Equivariance error ($\downarrow$)				Log-likelihood ($\uparrow$)
		Floor hue	Wall hue	Object hue	Scale	
VAE [14]	No (✗)	6924.63 ± 8.92	7746.37 ± 8.77	4383.54 ± 9.26	2609.59 ± 7.41	−11784.69 ± 4.87
β-VAE [12]	No (✗)	2243.95 ± 12.48	2279.23 ± 13.97	2188.73 ± 12.61	2037.94 ± 11.72	−11924.83 ± 5.64
FactorVAE [13]	No (✗)	1985.75 ± 13.26	1876.41 ± 11.93	1902.83 ± 12.27	1657.32 ± 11.05	−11802.17 ± 5.69
SlowVAE [9]	Weak (✓)	1247.36 ± 12.49	1314.86 ± 11.41	1102.28 ± 12.17	1058.74 ± 10.96	−11674.89 ± 5.74
TVAE [10]	Yes (✓)	1225.47 ± 9.82	1246.32 ± 9.54	1261.79 ± 9.86	1142.01 ± 9.37	−11475.48 ± 5.18
PoFlow [11]	Yes (✓)	885.46 ± 10.37	916.71 ± 10.49	912.48 ± 9.86	924.39 ± 10.05	−11335.84 ± 4.95
Ours	Yes (✓)	**613.29 ± 8.93**	**653.45 ± 9.48**	**605.79 ± 8.63**	**599.71 ± 9.34**	**−11215.42 ± 5.71**
Ours	Weak (✓)	690.84 ± 9.57	717.74 ± 10.65	681.59 ± 9.02	653.58 ± 9.57	−11279.61 ± 5.89

Table 6.3 VP Scores (%) on MNIST

Training set split (%)	Ours	PoFlow	TVAE	FactorVAE	β-VAE
10	**95.69**	93.05	89.91	85.92	87.31
1	**92.71**	91.27	88.15	84.46	85.25

Table 6.4 VP scores (%) on Shapes3D

Training set split (%)	Ours	PoFlow	TVAE	FactorVAE	β-VAE
10	**95.92**	91.48	88.27	84.49	85.91
1	**77.03**	72.32	68.39	63.83	65.78

6.4.3 Results on Complex Real-World Datasets

Tables 6.5 and 6.6 compare the equivariance error of our methods and the representative baselines on Falcol3D and Isaac3D, respectively. Notice that the values are much larger than previous datasets due to the increased image resolution. Our method still outperforms other baselines by a large margin and achieves reasonable equivariance error. Figure 6.5 displays the qualitative comparisons of our method against other baselines. Our method can precisely control the image transformations through our latent flows. *Overall, the above results demonstrate that our method can go beyond the toy setting and can be further applied to more complex real-world scenarios.*

6.5 Discussions

6.5.1 Extrapolation to Switching/Superposing Transformation

Extrapolation: switching transformations. In Fig. 6.6 we demonstrate that, empowered by our method, it is possible to switch latent transformation categories mid-way through the latent evolution and maintain coherence. That is, we perform $z_t = z_{t-1} + \nabla_z u^k$ for $t \leq T/2$ and then change to $z_t = z_{t-1} + \nabla_z u^j$ where $j \neq k$ for $t > T/2$. As can be seen, the factor of variation immediately changes after the transformation type is switched. Moreover, the transition phase is smooth and no other attributes of the image are influenced.

Extrapolation: superposing transformations. Besides switching transformations, our method also supports applying different transformations simultaneously, i.e., consistently performing $z_t = z_{t-1} + \sum_k^K \nabla_z u^k$ during the latent flow process. Figure 6.7 presents such exemplary visualizations of superposing two and all transformations simultaneously. In each case, the latent evolution corresponds to simultaneous smooth variations of multiple image

6.5 Discussions

Table 6.5 Equivariance error (↓) on Falcol3D [8]

Methods	Lighting intensity	Lighting X-dir	Lighting Y-dir	Lighting Z-dir	Camera X-pos	Camera Y-pos	Camera Y-pos
TVAE [10]	11477.81	12568.32	11807.34	11829.33	11539.69	11736.78	11951.45
PoFlow [11]	8312.97	7956.18	8519.39	8871.62	8116.82	8534.91	8994.63
Ours	5798.42	6145.09	6334.87	6782.84	6312.95	6513.68	6614.27

Table 6.6 Equivariance error ($\downarrow$) on Isaac3D [8]

Methods	Robot X-move	Robot Y-move	Camera height	Object scale	Lighting intensity	Lighting Y-dir	Object color	Wall color
TVAE [10]	8441.65	8348.23	8495.31	8251.34	8291.70	8741.07	8456.78	8512.09
PoFlow [11]	6572.19	6489.35	6319.82	6188.59	6517.40	6712.06	7056.98	6343.76
Ours	3659.72	3993.33	4170.27	4359.78	4225.34	4019.84	5514.97	3876.01

6.5 Discussions

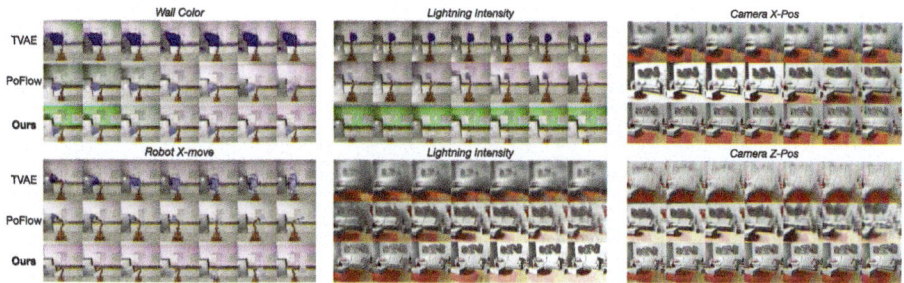

Fig. 6.5 Qualitative comparison of our method against TVAE and PoFlow on Falcol3D and Isaac3D

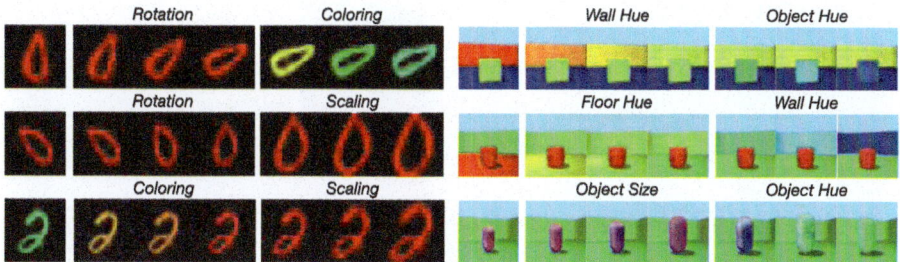

Fig. 6.6 Exemplary visualization of switching flow fields during the latent sample evolution

attributes. This indicates that our method also generalizes well to superposing different transformations.

Notice that we only apply single and separate transformations in the training stage. Switching or superposing transformations in the test phase can be thus understood as an extrapolation test to measure the generalization ability of the learned equivariance to novel compositions.

6.5.2 Equivariance Generalization to New Data

We also test whether the learned equivariance holds for Out-of-Distribution (OoD) data. To verify this, we validate our method on a test dataset that is different from the training set and therefore unseen to the model. Figure 6.8 displays the exemplary visualization results of the VAE trained on MNIST [6] but evaluated on dSprites [16]. Although the reconstruction quality is poor, the learned equivariance is still clearly effective as each transformation still operates as expected: scaling, rotation, and coloring transformations from top to bottom respectively.

This chapter presents a probabilistic sequential VAE that learns disentangled and approximately equivariant representations. Despite the efficacy in modeling diverse transformations, this method requires the supervision of pure transformation sequences, which may not apply

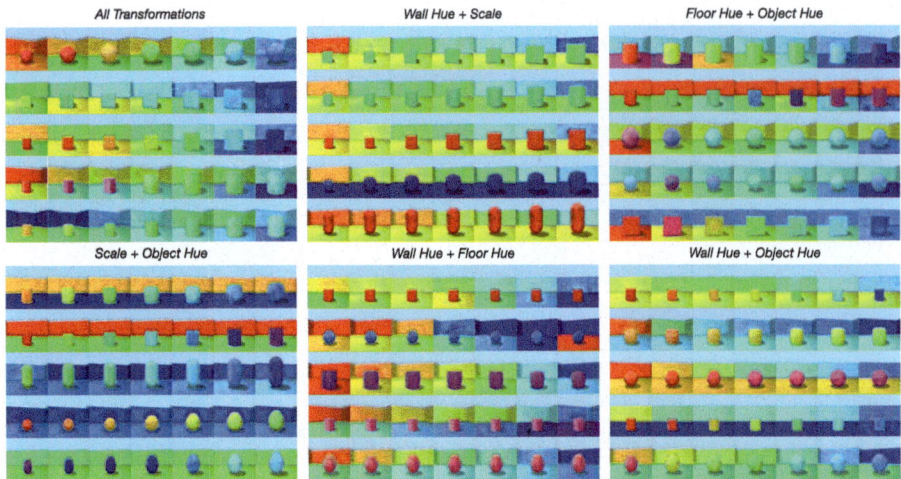

Fig. 6.7 Examples of combining flow fields simultaneously during latent evolution

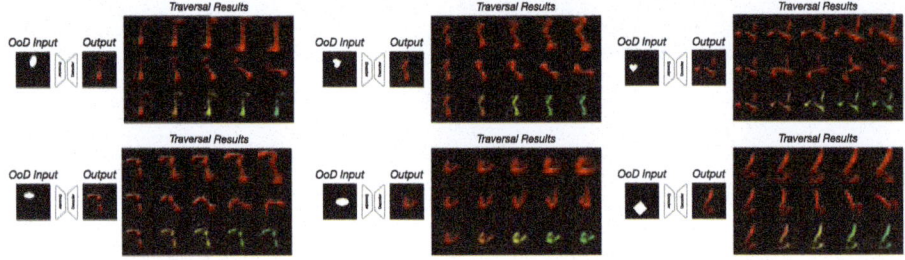

Fig. 6.8 Equivariance generalization to unseen OoD input data. Here the model is trained on MNIST [6] but the latent flow is tested on dSprites [16]

to real-world videos. In the next chapter, we will target this drawback and propose a sparsity-induced unsupervised representation learning approach based on natural video statistics. As we will demonstrate, removing the supervision requirement greatly facilitates more practical applications in real-world video understanding.

References

1. Jean-David Benamou and Yann Brenier. A computational fluid mechanics solution to the monge-kantorovich mass transfer problem. *Numerische Mathematik*, 84(3):375–393, 2000.
2. Danilo Rezende and Shakir Mohamed. Variational inference with normalizing flows. In *ICML*. PMLR, 2015.
3. Ricky T. Q. Chen, Yulia Rubanova, Jesse Bettencourt, and David Duvenaud. Neural ordinary differential equations. *NeurIPS*, 2019.

4. Eric Jang, Shixiang Gu, and Ben Poole. Categorical reparameterization with gumbel-softmax. *ICLR*, 2017.
5. M. Raissi, P. Perdikaris, and G.E. Karniadakis. Physics-informed neural networks: A deep learning framework for solving forward and inverse problems involving nonlinear partial differential equations. *Journal of Computational Physics*, 2019.
6. Yann LeCun. The mnist database of handwritten digits. 1998.
7. Chris Burgess and Hyunjik Kim. 3d shapes dataset. https://github.com/deepmind/3dshapes-dataset/, 2018.
8. Weili Nie, Tero Karras, Animesh Garg, Shoubhik Debnath, Anjul Patney, Ankit B Patel, and Anima Anandkumar. Semi-supervised stylegan for disentanglement learning. In *ICML*, 2020.
9. David Klindt, Lukas Schott, Yash Sharma, Ivan Ustyuzhaninov, Wieland Brendel, Matthias Bethge, and Dylan Paiton. Towards nonlinear disentanglement in natural data with temporal sparse coding. *ICLR*, 2021.
10. T Anderson Keller and Max Welling. Topographic vaes learn equivariant capsules. *NeurIPS*, 2021.
11. Yue Song, Andy Keller, Nicu Sebe, and Max Welling. Latent traversals in generative models as potential flows. In *ICML*. PMLR, 2023.
12. Irina Higgins, Loic Matthey, Arka Pal, Christopher Burgess, Xavier Glorot, Matthew Botvinick, Shakir Mohamed, and Alexander Lerchner. beta-vae: Learning basic visual concepts with a constrained variational framework. *ICLR*, 2016.
13. Hyunjik Kim and Andriy Mnih. Disentangling by factorising. In *ICML*, 2018.
14. Diederik P Kingma and Max Welling. Auto-encoding variational bayes. *ICLR*, 2014.
15. Xinqi Zhu, Chang Xu, and Dacheng Tao. Learning disentangled representations with latent variation predictability. In *ECCV*, 2020.
16. Loic Matthey, Irina Higgins, Demis Hassabis, and Alexander Lerchner. dsprites: Disentanglement testing sprites dataset, 2017.

Unsupervised Factorized Representation Learning Based on Sparse Transformation Analysis

7.1 Motivations

In this section, we outline the two important motivations for our sparsity-induced Helmholtz flow variational autoencoder.

7.1.1 Sparsity in Natural Videos

One early line of work in representation learning focused on ideas of redundancy reduction, believing that biological neural systems would naturally strive for an efficient code due to competitive pressures [1]. Building on this idea, the principles of sparsity and statistical independence of coding dimensions emerged as guidelines for learning such maximally efficient codes, eventually resulting in the frameworks of sparse coding [2] and independent component analysis [3]. Inspired by the fact that natural intelligence is embedded in a world where physical laws restrict observations to sequences of smooth transformations, these ideas of efficiency and sparsity were extended to include temporal dimensions. A seminal example is Slow Feature Analysis [4], a learning framework which assumes that individual latent variables are likely to change slowly over time. Models adhering to these principles were shown to learn invariances directly from data and uncover underlying generative factors if those factors had similar slow dynamics. Recent work has further shown that natural videos follow a specific sparse transition structure, meaning that the set of generative factors which describe a given input sequence is mostly constant over time with sparse transitions between which factors are active. Klindt et al. [5] then demonstrated that by building a model which incorporates this structure into its prior, it is possible to provably learn the true generative factors of video data in an unsupervised manner. While differing in implementation and methodology, all these frameworks appear in some sense to share the goals of learning meaningful, interpretable, 'disentangled', and controllable latent codes such that specific directions in the latent space corresponded to the independent factors which were responsible for generating the input data distribution.

7.1.2 Helmholtz Decomposition

The Helmholtz decomposition [6, 7] splits a sufficiently smooth vector field into two distinct components: a gradient field and a rotation field. The gradient field is defined as the gradient of some scalar potential function, which has vanishing curl and is thus irrotational. This vector field is often used for modeling translations. On the other hand, the rotation field is divergence-free and points in the transverse direction at each position. These two components constitute the Helmholtz decomposition that can model general vector fields. Our flow fields defined in Chap. 6 only adopt gradient flows in the latent space, which cannot model periodic transformations like rotation. In this chapter we further leverage the Helmholtz decomposition to obtain more expressive flow fields which can model invariant parts of the transformations.

7.2 The Generative Model

This section introduces the probabilistic framework of our generative model. We start with the factorization of sequence distributions, followed by the spike and slab priors, and end with the time evolution of the latent priors.

7.2.1 Factorized Sequence Distributions

Figure 7.1 depicts the plate diagram of our model through solid lines. As can be seen, our model defines a distribution over N sequences of observed variables $\bar{x} = \{x_0, x_1 \ldots, x_T\}$. The sequence distribution is factorized into K distinct basic components as we assume each observed sequence is generated by the linear combination of K separate basis flows in latent space. To model the discrete sequences of observations, we aim to define a joint distribution with a similarly discrete sequence of latent variables $\bar{z} = \{z_0, z_1 \ldots, z_T\}$ describing the observations, and $\bar{g} = \{y_1 \cdot \tilde{g}_1, y_2 \cdot \tilde{g}_2 \ldots, y_T \cdot \tilde{g}_T\}$ describing the transformation *type* (y_t) and *speed* ($\tilde{g}_t$) happening between neighboring observations. Specifically, we assert the following factorization of the joint distribution over T timesteps:

$$p(\bar{x}, \bar{z}, \bar{g}) = p(g_1) p(z_0) p(x_0 | z_0) \\ \prod_{t=1}^{T} p(z_t | z_{t-1}, g_t) p(x_t | z_t, g_t) p(g_{t+1} | g_t). \tag{7.1}$$

7.2 The Generative Model

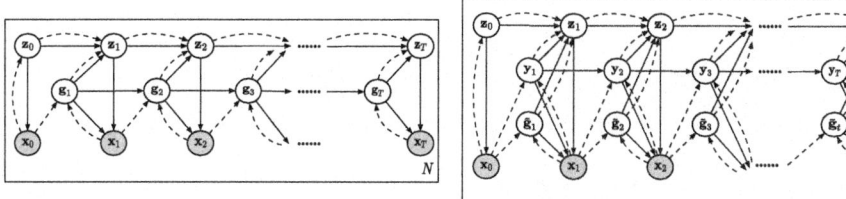

Fig. 7.1 Our model across N sequences in plate notation (Left) and a detailed version with decomposed spike and slab components (Right). White nodes denote latent variables, shaded nodes denote observed variables, solid lines denote the generative model, and dashed lines denote the approximate posterior. Different from the spike component y_t, the slab variable $\tilde{g}_t$ is independent across timesteps

Here $p(z_0)$ is a standard Normal distribution, $p(x_t | z_t, g_t)$ asserts a mapping from latents to observations, and $p(\bar{g})$ is the sequence of the random variables that controls the temporal variations of the transformation type and speed.

7.2.2 Spike and Slab Priors

We model real-world video as a sparse combination of transformation primitives. To model this transition sparsity, we impose a spike and slab prior [8] on the transformation variable g_t for generating the sequences. The distribution is factorized as follows:

$$p(g_t) = p(y_t) p(\tilde{g}_t) \tag{7.2}$$

where the 'spike' variable y_t is a multi-hot vector that selects the specific transformation primitives to combine, and the 'slab' variable $\tilde{g}_t$ controls the transformation speed. The spike component usually concentrates its mass around zero, whereas the slab component is spread over a range of plausible values (e.g. Gaussian or Laplace distributions). Their product $y_t \cdot \tilde{g}_t$ allows shrinking some values of $\tilde{g}_t$ to zero and therefore effectively promotes sparsity. We further factorize the joint distribution of these variables over time as:

$$p(\bar{g}) = p(y_1) \prod_{t=2}^{T} p(y_t | y_{t-1}) \prod_{t=1}^{T} p(\tilde{g}_t) \tag{7.3}$$

Here the conditional update $p(y_t | y_{t-1})$ is enforced to ensure that the transformation type is temporally coherent and varies sparsely. We do not enforce such constraints to $p(\tilde{g}_t)$ as the Laplace distribution is very concentrated around the center and is already sparsity-inducing.
Spike Priors. For the spike variable, we define the following multivariate Bernoulli prior:

$$\begin{aligned} p(y_1) &= \mathrm{Ber}(P_1), \\ p(y_t | y_{t-1}) &= \mathrm{Ber}(\sigma(a + b y_{t-1})). \end{aligned} \tag{7.4}$$

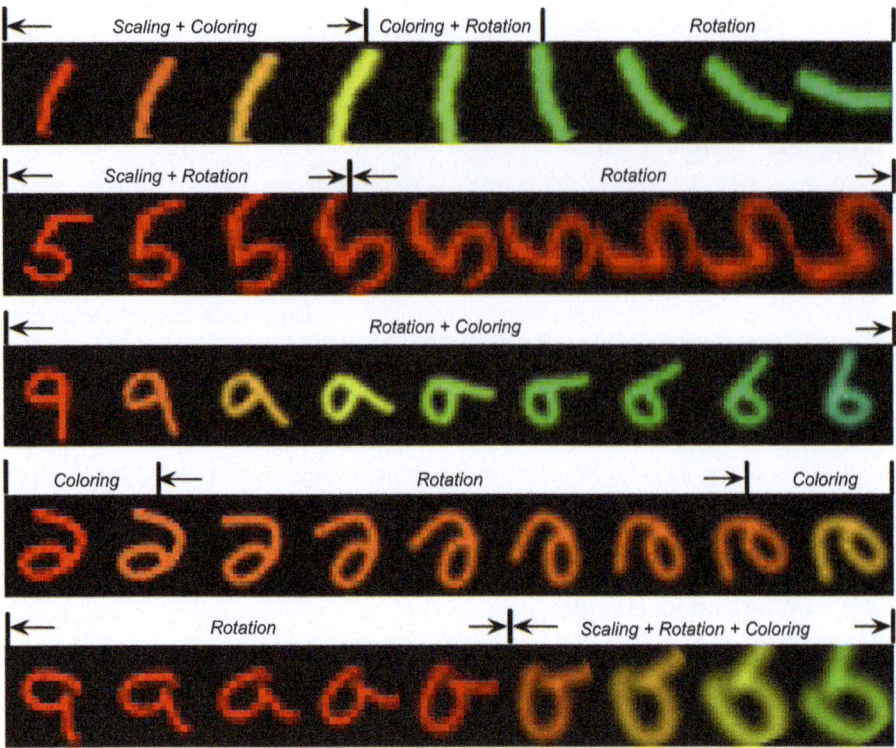

Fig. 7.2 Exemplary sequences generated by our spike prior

where P_1 is the probability of switching on, $\sigma(\cdot)$ denotes the activation function, and a, b are hyper-parameters that determine the transition probability. Since we aim to obtain data sequences with smooth variations, the temporal transitions of y_t need to be sparse. This is achieved by setting $\sigma(a)$ to be low and $\sigma(a + b)$ to be high. When drawing samples from the Bernoulli distributions in Eq. (7.4), we reject all-zero samples to avoid generating sequences where no single transformations are applied. Figure 7.2 displays the generated sequences of MNIST [9] using spike priors. The variations align with natural videos – the transitions happen occasionally and smoothly.

Slab Priors. For the slab component, we use a Laplace distribution:

$$p(\tilde{g}_t) = \text{Laplace}(\mu, \lambda) = \frac{1}{2\lambda} \exp(-\frac{|\tilde{g}_t - \mu|}{\lambda}) \tag{7.5}$$

where μ is the mean, and λ is the scale parameter that controls the sharpness of the distribution. A sharper Laplace distribution will generate speeds more peaked around μ. In our experiments we set $\mu = 1$. The slab variable introduces the additional control of the transformation speed, which further mimics the dynamics of real-world videos.

7.2.3 Prior Time Evolution

In line with our flow-factorized VAE Chap. 6, as we do not assume any prior knowledge of each transformation, we would like to enforce minimally informative priors. This can be achieved by considering the time evolution as Brownian motion, i.e., random trajectories. To this end, we define the potential function $\psi^k(z) = -D_k \log p(z_t)$ which advects the density $p(z)$ through the induced velocity field $\nabla \psi^k(z)$. Then according to the continuity equation, the prior evolves as:

$$\partial_t p(z_t) = -\nabla \cdot \left(p(z_t) v(z) \right) = \sum_k (g_t^k D_k) \nabla^2 p(z_t) \quad (7.6)$$

where D_k is a learnable constant coefficient which is distinct for each k. The time evolution of the prior distribution thus follows a weighted diffusion process.

7.3 Helmholtz Flow Variational Autoencoders

In this section, we first introduce the Helmholtz decomposition of the latent flow fields, then proceed to explain the inference over observed variables. Finally, we detail the time evolution of our latent prior and posterior.

7.3.1 Helmholtz Decomposed Latent Flows

By the Helmholtz decomposition [6, 7, 10], a vector field $\mathbf{F}$ can be uniquely represented by the sum of two vector fields such that:

$$\begin{aligned} \mathbf{F}(\mathbf{x}) &= \mathbf{G}(\mathbf{x}) + \mathbf{R}(\mathbf{x}) \\ \mathbf{G}(\mathbf{x}) &= -\nabla \Phi(\mathbf{x}), \quad \nabla \cdot \mathbf{R}(\mathbf{x}) = \mathbf{0} \end{aligned} \quad (7.7)$$

where $\mathbf{G}(\mathbf{x})$ is the irrotational (curl-free) component ($\nabla \times \mathbf{G}(x) = 0$), and $\mathbf{R}(\mathbf{x})$ is the divergence-free component. We then model the latent evolution using $\mathbf{F}$ as:

$$\begin{aligned} z_t &= z_{t-1} + \sum_k g_t^k \mathbf{F}^k(z) \\ &= z_{t-1} + \sum_k \tilde{g}_t^k y_t^k \left(\nabla u^k(z, t) + r^k(z) \right) \end{aligned} \quad (7.8)$$

where $u(z, t) = \text{MLP}(z; t) \in \mathbb{R}^1$ parameterizes the scalar spatiotemporal potential, and $r(z) = \text{MLP}(z) \in \mathbb{R}^d$ defines the divergence-free vector field. We achieve this divergence-free constraint by imposing the following PINN loss:

$$\mathcal{L}_{DIV} = \frac{1}{T} \sum_t \sum_k \left(g_t^k \nabla \cdot \boldsymbol{r}^k(z_t) \right)^2 \tag{7.9}$$

Richter et al. [11] proposed an approach to construct strict divergence-free vector fields. However, it requires computing the full Jacobian matrix at every step, which is memory-intensive and computationally slow. For faster computation, we use a PINN to approximate the vector field. Compared with our previous works which only include the curl-free component **G**, this parameterization allows for significantly increased flexibility in modeling periodic dynamics in the latent space. Furthermore, as will be illustrated later in Sect. 7.5, we expect that our model automatically learns to segregate periodic and non-periodic transformations into these two components.

7.3.2 Evidence Lower Bound and Inference

We define the approximate posterior of the transformation variable $\boldsymbol{g}_t$ to factorize as follows:

$$q_\gamma(\bar{\boldsymbol{g}}|\bar{\boldsymbol{x}}) = \prod_{t=1}^{T} q(\boldsymbol{y}_t|\boldsymbol{x}_t, \boldsymbol{x}_{t-1}) q(\tilde{\boldsymbol{g}}_t|\boldsymbol{x}_t, \boldsymbol{x}_{t-1}) \tag{7.10}$$

Both the spike and slab variables are inferred from the neighboring images. For the latent particles, we have the following factorization of the approximate posterior:

$$q_\theta(\bar{\boldsymbol{z}}|\bar{\boldsymbol{x}}, \bar{\boldsymbol{g}}) = q(z_0|\boldsymbol{x}_0) \prod_{t=1}^{T} q(z_t|z_{t-1}, \boldsymbol{g}_t) \tag{7.11}$$

In essence, given the transformation coefficient $\bar{\boldsymbol{g}}$, our posterior only considers information from $\boldsymbol{x}_0$ instead of the full sequence. However, as can be seen from Eq. (7.10), each $\boldsymbol{g}_t$ can see the variations happening between $\boldsymbol{x}_t$ and $\boldsymbol{x}_{t-1}$, and thus $\bar{\boldsymbol{g}}$ contains the remaining sequence information.

We derive the lower bound to model evidence (ELBO) as:

$$\begin{aligned}
\log p(\bar{\boldsymbol{x}}) &= \mathbb{e}_{q_\theta(\bar{z}|\bar{x},\bar{g}), q_\gamma(\bar{g}|\bar{x})} \left[\log \frac{p(\bar{\boldsymbol{x}}, \bar{z}, \bar{\boldsymbol{g}})}{q(\bar{z}, \bar{\boldsymbol{g}}|\bar{\boldsymbol{x}})} \frac{q(\bar{z}|\bar{\boldsymbol{x}}, \bar{\boldsymbol{g}})}{p(\bar{z}|\bar{\boldsymbol{x}}, \bar{\boldsymbol{g}})} \right] \\
&\geq \mathbb{e}_{q_\theta(\bar{z}|\bar{x},\bar{g}), q_\gamma(\bar{g}|\bar{x})} \left[\log \frac{p(\bar{\boldsymbol{x}}|\bar{z}, \bar{\boldsymbol{g}}) p(\bar{z}|\bar{\boldsymbol{g}})}{q(\bar{z}|\bar{\boldsymbol{x}}, \bar{\boldsymbol{g}})} \frac{p(\bar{\boldsymbol{g}})}{q(\bar{\boldsymbol{g}}|\bar{\boldsymbol{x}})} \right] \\
&= \mathbb{e}_{q_\theta(\bar{z}|\bar{x},\bar{g})} \left[\log p(\bar{\boldsymbol{x}}|\bar{z}, \bar{\boldsymbol{g}}) \right] + \mathbb{e}_{q_\theta(\bar{z}|\bar{x},\bar{g})} \left[\log \frac{p(\bar{z}|\bar{\boldsymbol{g}})}{q(\bar{z}|\bar{\boldsymbol{x}}, \bar{\boldsymbol{g}})} \right] \\
&\quad + \mathbb{e}_{q_\gamma(\bar{g}|\bar{x})} \left[\log \frac{p(\bar{\boldsymbol{g}})}{q(\bar{\boldsymbol{g}}|\bar{\boldsymbol{x}})} \right]
\end{aligned} \tag{7.12}$$

7.3 Helmholtz Flow Variational Autoencoders

The above ELBO can be further re-written as:

$$\log p(\tilde{x}) \geq \sum_{t=0}^{T} e_{q_\theta(\tilde{z}|\tilde{x},\tilde{g})} \big[\log p(x_t|z_t, g_{t+1}) \big]$$
$$- e_{q_\theta(\tilde{z}|k)} \big[D_{\text{KL}} [q_\theta(z_0|x_0) || p(z_0)] \big]$$
$$- \sum_{t=1}^{T} e_{q_\theta(\tilde{z}|\tilde{x},\tilde{g})} \big[D_{\text{KL}} [q_\theta(z_t|z_{t-1}, g_t) || p(z_t|z_{t-1}, g_t)] \big]$$
$$- e_{q_\gamma(\tilde{g}|\tilde{x})} \big[D_{\text{KL}} [q_\gamma(y_1|x_1, x_0) || p(y_1)] \big] \qquad (7.13)$$
$$- \sum_{t=2}^{T} e_{q_\gamma(\tilde{g}|\tilde{x})} \big[D_{\text{KL}} [q_\gamma(y_t|x_t, x_{t-1}) || p(y_t|y_{t-1})] \big]$$
$$- \sum_{t=1}^{T} e_{q_\gamma(\tilde{g}|\tilde{x})} \big[D_{\text{KL}} [q_\gamma(\tilde{g}_t|x_t, x_{t-1}) || p(\tilde{g}_t)] \big]$$

Compared with the objective of a traditional VAE, our model additionally involves the time evolution of the priors and posteriors. As noted in Sect. 7.2.2, we set $p(\tilde{g}_t)$ to follow a Laplace distribution and impose multivariate Bernoulli distributions to $p(y_1)$ and $p(y_t|y_{t-1})$. The KL divergence on $q_\gamma(y_t|x_t, x_{t-1})$ serves as regularization to encourage the sparsity of y_t. That being said, the posterior $q_\gamma(y_t|x_t, x_{t-1})$ learns to model the transformations using as few vector fields as possible, which naturally disentangles the input variations into distinct flow fields. We apply the Gumbel-Sigmoid trick [12] for the re-parameterization and sampling of the posterior $q_\gamma(y_t|x_t, x_{t-1})$.

7.3.3 Posterior Time Evolution

We use the change of variables formula again to derive the conditional update of $q(z_t|z_{t-1}, g_t)$. Given the function of the sample evolution $z_t = h(z_{t-1}, g_t) = z_{t-1} + \sum_k g_t^k (\nabla_z u^k + r^k)$, we still have the relation:

$$q(z_t|z_{t-1}, g_t) = q(z_{t-1}) \left| \frac{dh(z_{t-1}, g_t)}{dz_{t-1}} \right|^{-1} \qquad (7.14)$$

Discretizing the continuous form and taking the logarithm yields the normalizing-flow-like density evolution:

$$\log q(z_t|z_{t-1}, g_t) = \log q(z_{t-1})$$
$$- \log |I + \sum_k g_t^k (\nabla \nabla^T u^k + \nabla (r^k)^T)|$$
$$\approx \log q(z_{t-1}) - \sum_k g_t^k (\nabla^2 u^k + \nabla \cdot r^k)) \qquad (7.15)$$
$$= \log q(z_{t-1}) \sum_k g_t^k \nabla^2 u^k$$

where we take a Taylor approximation to expand the probability update term and have $\nabla \cdot r^k = 0$ by construction. We therefore expect the determinant of ∇r^k to be very small and hardly influence the density evolution. It is thus sufficient to not account for the impact of r^k here.

Algorithm 7.1 Training algorithm of our method.

Require: Encoder m, maximum traversal step T, image transform function n, and posteriors q_θ, q_γ.
1: **repeat**
2:    Encode: $z_0 = m(x_0)$
3:    Traversal Step Counter: $i = 0$
4:    **while** $i \leq T$ **do**
5:      Sample: $g_{i+1} \sim p(g_{i+1})$
6:      Image transform: $x_{i+1} = n(x_i, g_{i+1})$
7:      Infer: $\hat{g}_{i+1} = q_\gamma([x_i; x_{i+1}])$
8:      Flow: $z_{i+1} = z_i + \sum \hat{g}_{i+1}^k (\nabla u^k(z,t) + r^k(z))$
9:      Decode: $x_{i+1} = q_\theta(z_{i+1})$
10:     $i = i + 1$
11:   **end while**
12:   Optimize the ELBO $\log p(\bar{x})$ in Eq. (7.12) and the PINN losses $\mathcal{L}_{DIV}$ and $\mathcal{L}_{HJ}$.
13: **until** converged

7.4 Experiments

7.4.1 Evaluation Methods

Datasets. We evaluate our method on two widely-used benchmarks for standard representation learning, namely MNIST [9] and Shapes3D [13]. For MNIST, The basic 'pure' transformations consist of Scaling, Rotation, and Coloring. For Shapes3D, we use the self-contained four transformation primitives, including Floor Hue, Wall Hue, Object Hue, and Scale. On both datasets, we use our spike and slab prior to generate sequences that are composed of 'composite' transformations.

Beyond the toy datasets, we also evaluate our method on challenging Falcol3D and Issac3D [14], two complex large-scale and real-world datasets that contain sequences of different transformations. Specifically, Falcol3D consists of indoor 3D scenes with different lighting conditions and camera positions, while Isaac3D is comprised of various robot-arm movements in dynamic environments. Since the image sequences are short, we do not consider speed variations but only enforce the spike prior to generate data sequences with sparsely-varying transformations.

7.4 Experiments

We further conduct some preliminary experiments of applying our method to real-world video analysis, including autonomous driving videos on Cityscape [15] and behavior videos of social agents on CalMS [16]. Different from the used datasets above, we directly feed raw video sequences as input and let the model discover independent motions.

Baselines. We compare our method with some representative approaches in the field of disentangled and equivariant representation learning, including LatentFlow [17] and PoFlow [18] introduced in Chaps. 5 and 6 which adopt potential flow to evolve the latent samples, Topographic VAE (TVAE) [19] which posses topographic structured latent space, SlowVAE [5] which proposes the sparse Laplacian prior $p(z_t|z_{t-1}) = \prod \alpha\lambda/2\Gamma(1/\alpha) \exp(-\lambda|z_{t,i} - z_{t-1,i}|^\alpha)$, and β-VAE [20] and FactorVAE [21] which encourage the factorization of the single dimensions of latent samples. We also use the vanilla VAE [22] as a controlled baseline.

Metrics. We mainly evaluate the baselines using the equivariance error which is defined as $\mathcal{E}_k = \sum_{t=1}^{T} |x_t - \texttt{Decode}(z_t)|$ where x_t is the element of sequences of each transformation primitive (e.g., scaling and rotation). Since our method is unsupervised, we inspect the traversal results of each basic vector field $\nabla u^k + r^k$ and select the index whose flow looks the most like the target transformation. The average log-likelihood of the sequence is also evaluated on the test set. Besides these two metrics, we also adopt the metric Variational Predictability (VP) score [23] to evaluate the disentanglement performance.

7.4.2 Learning Composable Equivariant Latent Flows

Qualitative Results. Figure 7.3 displays the traversal results of each learned latent flow under different speeds on MNIST. Our model simultaneously disentangles the transformation categories and speeds into these vector fields in an unsupervised manner. We see that each flow field corresponds to a distinct transformation and further presents a precise control of the transformation speed. When increasing the magnitude of g^k, the transformation process

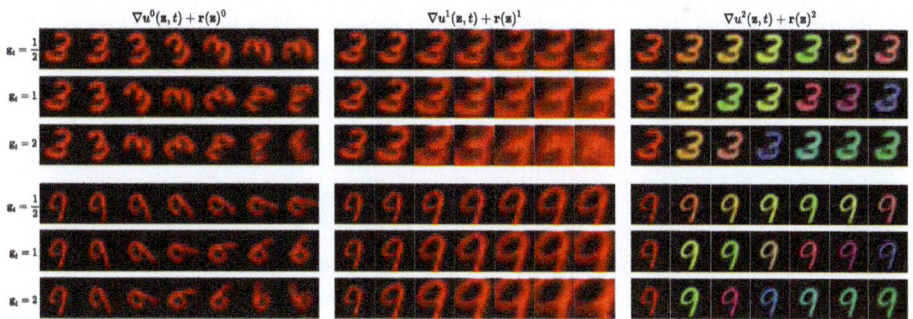

Fig. 7.3 Traversals using individual learned flows $k=\{0, 1, 2\}$ from left to right with speeds $g_t = \{\frac{1}{2}, 1, 2\}$ from top to bottom

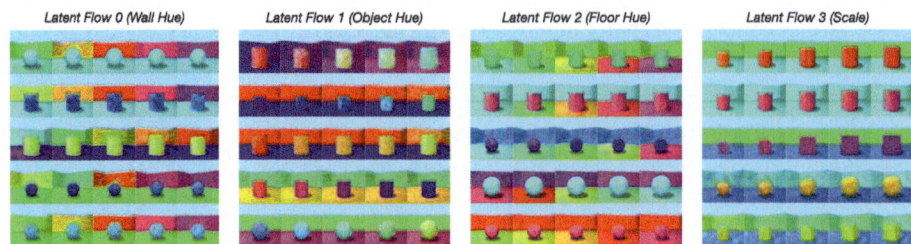

Fig. 7.4 Traversals using each individual learned flow field on Shapes3D [13]. In the bracket, we indicate the transformation which the traversal results look most like. Each latent flow has separate samples per row transforming from left to right

Fig. 7.5 Traversals using learned flows with different speeds $g_t = \{\frac{1}{2}, 1\}$ on Shapes3D

will be accelerated, i.e., the object will rotate more degrees, get scaled with a larger factor, and change the hue more. Figures 7.4 and 7.5 present the traversal results and the speed variations on Shapes3D. Our method still allows for disentanglement of the transformation categories and speed. We note that speed control is a major merit of our approach as the explicit control of transformation speed is seldom explored in deep representation learning.
Disentanglement Scores. Table 7.2 and 7.3 present the quantitative evaluation of the VP scores with different split ratios of the training set on MNIST and Shapes3D. Surprisingly, our method outperforms all baselines, including both supervised and unsupervised ones. Different from the supervised methods where each vector field is forced to learn one transformation, our model naturally disentangles the transformations into these learned flows through sparsity. We suspect that this gap might make our flows easier to be distinguished by small neural networks.
Quantitative Results. Tables 7.1 and 7.4 present the evaluation results of the equivariance error and log-likelihood on MNIST and Shapes3D, respectively. We see our method achieves very competitive performance against other baselines. Specifically, our method outperforms all the unsupervised approaches by a large margin on equivariance error and rivals PoFlow [18] which requires supervision of each transformation primitive. Moreover, our method yields the highest log-likelihood on the test set, which is likely accounted for by the fact that our method incorporates a sophisticated transformation-centric prior over

7.4 Experiments

Table 7.1 Equivariance error $\mathcal{E}_k$ and average log-likelihood $\log p(x_t)$ on MNIST [9]

Methods	Supervision?	Equivariance error ($\downarrow$)			Log-likelihood ($\uparrow$)
		Scaling	Rotation	Coloring	
VAE [22]	No (✗)	1275.31 ± 1.89	1310.72 ± 2.19	1368.92 ± 2.33	−2206.17 ± 1.83
β-VAE [20]	No (✗)	741.58 ± 4.57	751.32 ± 5.22	808.16 ± 5.03	−2224.67 ± 2.35
FactorVAE [21]	No (✗)	659.71 ± 4.89	632.44 ± 5.76	662.18 ± 5.26	−2209.33 ± 2.47
SlowVAE [5]	Weak ( )	461.59 ± 5.37	447.46 ± 5.46	398.12 ± 4.83	−2197.68 ± 2.39
TVAE [19]	Yes (✓)	505.19 ± 2.77	493.28 ± 3.37	451.25 ± 2.76	−2181.13 ± 1.87
PoFlow [18]	Yes (✓)	234.78 ± 2.91	231.42 ± 2.98	240.57 ± 2.58	−2145.03 ± 2.01
LatentFlow [17]	Yes (✓)	**185.42 ± 2.35**	**153.54 ± 3.10**	**158.57 ± 2.95**	−2112.45 ± 1.57
LatentFlow [17]	Weak ( )	193.84 ± 2.47	157.16 ± 3.24	165.19 ± 2.78	−2119.94 ± 1.76
Ours	No (✗)	281.32 ± 4.71	230.93 ± 5.02	292.85 ± 4.58	**−2107.65 ± 2.27**

Table 7.2 VP Scores (%) on MNIST

Training set (%)	Ours	LatentFlow	PoFlow	TVAE	FactorVAE
10	**98.85**	95.69	93.05	89.91	85.92
1	**97.04**	92.71	91.27	88.15	84.46

Table 7.3 VP Scores (%) on Shapes3D

Training set (%)	Ours	LatentFlow	PoFlow	TVAE	FactorVAE
10	**97.98**	95.92	91.48	88.27	84.49
1	**86.09**	77.03	72.32	68.39	63.83

latent states which matches the statistics of the data. The sparse combination of multiple transformations thus can be seen as a kind of data augmentation. Among the transformations, the rotation has the smallest equivariance error. We expect this gain is largely due to the rotational vector field $r(z)$ introduced by the Helmholtz decomposition.

Results on Composite Transformations. Besides the standard evaluation of individual transformations, it would be interesting to validate the equivariance property of composite transformations. To this end, we measure their equivariance error using the predicted spike and slab components g_t to combine different flow fields linearly. Table 7.5 compares the performance against two strong baselines. Since we explicitly superpose latent flows in the training, our unsupervised method outperforms these supervised approaches significantly, which further demonstrates the flexible linear composability of our latent flows.

Table 7.4 Equivariance error $\mathcal{E}_k$ and average log-likelihood $\log p(x_t)$ on Shapes3D [13]

Methods	Supervision?	Equivariance error (↓)				Log-likelihood (↑)
		Floor hue	Wall hue	Object hue	Scale	
VAE [22]	No (✗)	6924.63 ± 8.92	7746.37 ± 8.77	4383.54 ± 9.26	2609.59 ± 7.41	−11784.69 ± 4.87
β-VAE [20]	No (✗)	2243.95 ± 12.48	2279.23 ± 13.97	2188.73 ± 12.61	2037.94 ± 11.72	−11924.83 ± 5.64
FactorVAE [21]	No (✗)	1985.75 ± 13.26	1876.41 ± 11.93	1902.83 ± 12.27	1657.32 ± 11.05	−11802.17 ± 5.69
SlowVAE [5]	Weak (✓)	1247.36 ± 12.49	1314.86 ± 11.41	1102.28 ± 12.17	1058.74 ± 10.96	−11674.89 ± 5.74
TVAE [19]	Yes (✓)	1225.47 ± 9.82	1246.32 ± 9.54	1261.79 ± 9.86	1142.01 ± 9.37	−11475.48 ± 5.18
PoFlow [18]	Yes (✓)	885.46 ± 10.37	916.71 ± 10.49	912.48 ± 9.86	924.39 ± 10.05	−11335.84 ± 4.95
LatentFlow [17]	Yes (✓)	**613.29 ± 8.93**	**653.45 ± 9.48**	**605.79 ± 8.63**	**599.71 ± 9.34**	**−11215.42 ± 5.71**
LatentFlow [17]	Weak (✓)	690.84 ± 9.57	717.74 ± 10.65	681.59 ± 9.02	653.58 ± 9.57	−11279.61 ± 5.89
Ours	No (✗)	**1005.23 ± 11.79**	**1171.69 ± 13.64**	**928.10 ± 11.58**	**894.77 ± 10.94**	**−11199.93 ± 5.93**

7.4 Experiments

Table 7.5 Equivariance error $\mathcal{E}_k$ of composite transformations. For both baselines, we linearly combine their latent flows

Methods	Scaling + Rotation	Scaling + Coloring	Rotation + Coloring
PoFlow	582.17±4.33	597.20 ± 3.94	574.86 ± 4.07
LatentFlow	493.75 ± 3.62	501.82 ± 4.07	452.63 ± 3.29
Ours	**293.45 ± 4.12**	**321.82 ± 4.74**	**407.95 ± 4.58**

7.4.3 Analysis of Real-World Videos

Robot Arms and Indoor Scenes. Figures 7.6 and 7.7 show the learned latent flows on Falcol3D and Issac3D [14], respectively. As can be seen from the figures, even on these challenging large-scale datasets, our method still allows for unsupervised disentanglement of complex real-world transformations. Tables 7.6 and 7.7 compares the equivariance error on the two datasets. Similar to the results on MNIST and Shapes3D, our method achieves

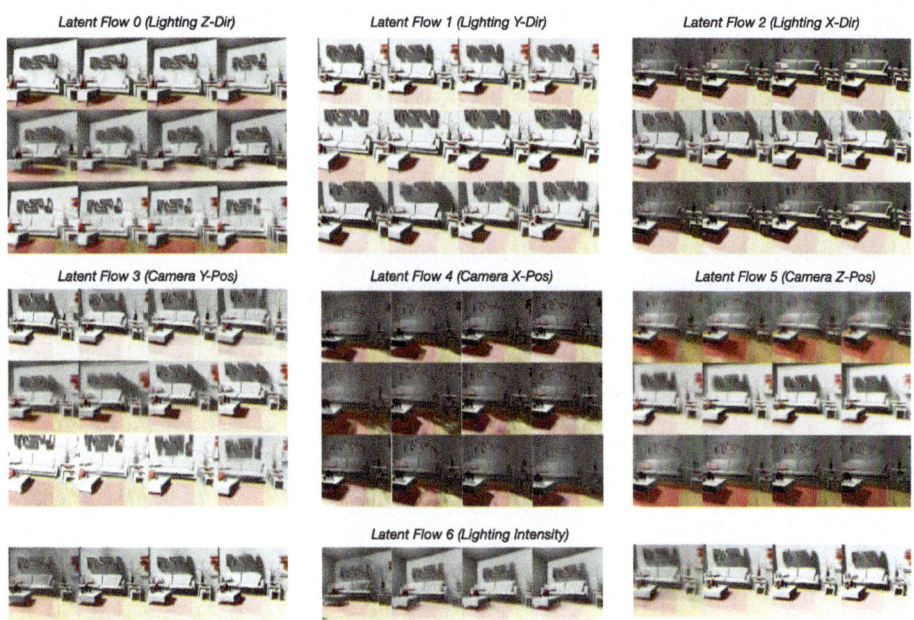

Fig. 7.6 Traversals using each individual learned flow field on Falcol3D [14]. In the bracket, we indicate the transformation which the traversal results look most like. Each latent flow has separate samples per row transforming from left to right. The bottom row displays the traversal result generated by the 6'th latent flow field

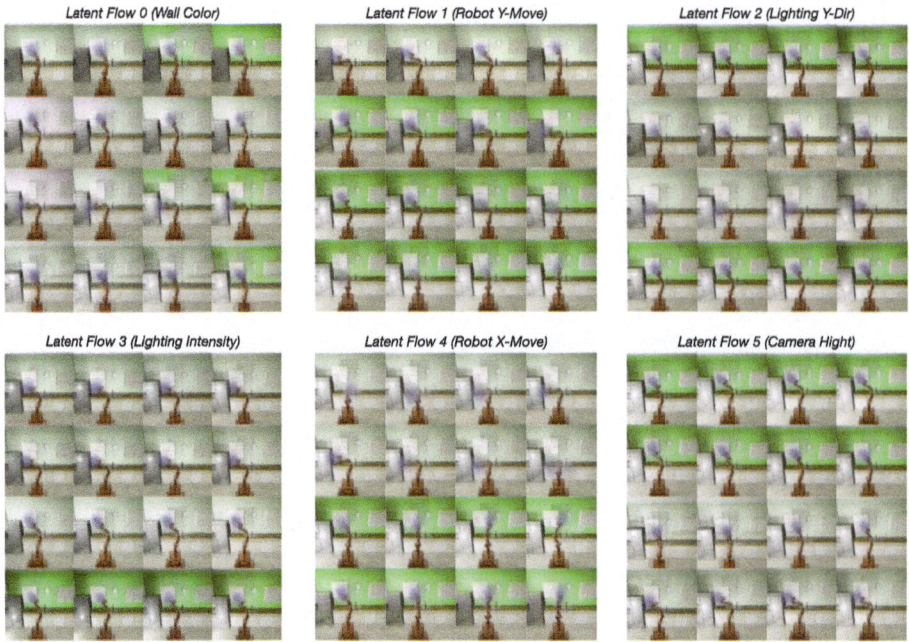

Fig. 7.7 Traversals using each individual learned flow field on Issac3D [14]. In the bracket, we indicate the transformation which the traversal results look most like. Each latent flow has separate samples per row transforming from left to right

very competitive performance against supervised ones. This demonstrates that the proposed sparsity priors also scale up to sequences of complex transformations.

Agent Behavioral Videos. We apply our method to disentangle the complex social interactions of mice on CalMS [16]. On this dataset, there exist three ground truth interactions, namely 'investigation', 'attack', and 'mount'. We thus define 3 latent flows and let the model learn the interactions from the raw videos. The images are of size 128×128.

Figure 7.8 displays the exemplary traversal results of three distinct latent flows. We could have reasonable interpretations of the interaction categories for these latent flows. Specifically, we might interpret latent flow 1 as 'investigation', latent flow 2 as 'attack', and latent flow 3 as 'mount', respectively. To validate if the interpretations align with human annotations, we compute the correlation between the predicted spike variable and the behavior labels. Table 7.8 reports the classification accuracy of each interaction class. As an unsupervised approach, our method achieves competitive results against the supervised baselines, indicating that the sparsity prior can help disentangle the mouse behaviors. Figure 7.9 displays a few examples of image sequences and the reconstruction results using the spike prior and the latent flow fields. Our method can reconstruct the behaviors that are close to the ground truth.

7.4 Experiments

Table 7.6 Equivariance error $\mathcal{E}_k$ on Falcor3D

Methods	Lighting intensity	Lighting X-dir	Lighting Y-dir	Lighting Z-dir	Camera X-pos	Camera Y-pos	Camera Z-pos
TVAE [19]	11477.81	12568.32	11807.34	11829.33	11539.69	11736.78	11951.45
PoFlow [18]	8312.97	7956.18	8519.39	8871.62	8116.82	8534.91	8994.63
LatentFlow [17]	5798.42	6145.09	6334.87	6782.84	6312.95	6513.68	6614.27
Ours	8672.91	8146.91	8729.06	9023.56	8064.75	8856.92	9134.02

Table 7.7 Equivariance error $\mathcal{E}_k$ on Issac3D

Methods	Robot X-move	Robot Y-move	Camera height	Object scale	Lighting intensity	Lighting Y-dir	Object color	Wall color
TVAE [19]	8441.65	8348.23	8495.31	8251.34	8291.70	8741.07	8456.78	8512.09
PoFlow [18]	6572.19	6489.35	6319.82	6188.59	6517.40	6712.06	7056.98	6343.76
LatentFlow [17]	3659.72	3993.33	4170.27	4359.78	4225.34	4019.84	5514.97	3876.01
Ours	7012.34	6399.57	6589.48	6104.74	6298.16	6517.23	6674.98	6519.38

7.4 Experiments

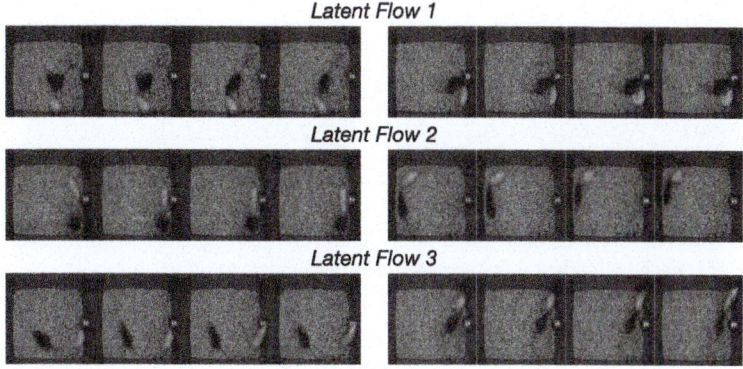

Fig. 7.8 Traversal results of learned latent flows on CalMS [16]. For each latent flow, we display two exemplary sequences, and the flow transforms the image from left to right

Table 7.8 Behavior classification results on CalMS [16]

Method	MARS [24]	B-Kind [16]	Trajectory-LSTM [16]	Ours
Supervision?	Yes (✓)	Yes (✓)	Yes (✓)	No (✗)
mAP	0.880	0.852	0.712	0.793

Fig. 7.9 Exemplary comparisons of the ground truth image sequences and reconstruction results. For each sequence, we start with reconstructing the initial frame and use the spike component and latent flow fields to generate the rest frames

Autonomous Driving Videos. Finally, we take a step further to evaluate our method on Cityscape [15], the challenging real-world autonomous driving videos. We take the sequences of segmentation masks as the training data and downsample the resolution to 64×64. Figure 7.10 displays some exemplary traversals of different latent flows. On this dataset, there are no ground truth generative factors so we may have some reasonable interpretations according to the disentangled transformations: we may interpret latent flow 0 as turning left (the sidewalk region on the right side shrinks), latent flow 1 as getting closer to the front car (the car region expands), latent flow 2 as getting away from the front car (the car region shrinks and disappears), and latent flow 3 as changing the right side from terrain to sidewalk. *Notice that this is an initial attempt to apply our method to complex*

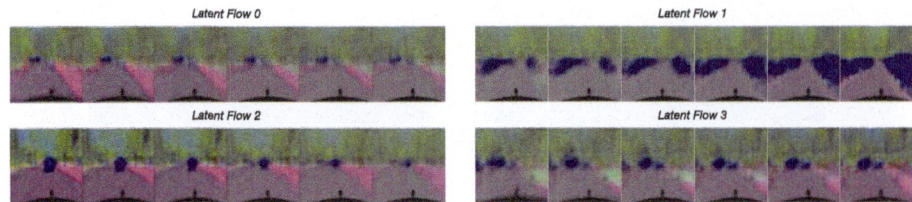

Fig. 7.10 Traversals results of learned flow fields on downsampled segmentation masks of Cityscape [15]. Each latent flow transforms the image from left to right

real-world video analysis. Nonetheless, this preliminary experiment demonstrates that our method could have real-world applicability for video understanding.

7.5 Discussions

7.5.1 Switchability and Composability

Figures 7.11 and 7.12 display the traversal results of switching and combining different latent flows, respectively. Our model is able to switch to another vector field primitive with smooth output transitions and also supports performing multiple transformations simultaneously. This result indicates that our approach allows for flexible generalization to switchability and linear composability of arbitrary latent flows.

7.5.2 Handling Periodic Transformations

Figure 7.13 compares the traversal results of two latent flows using different types of vector fields. For rotation, the divergence-free vector field r^0 dominates this transformation whereas the curl-free vector field ∇u^0 has little impact. This meets our expectation that periodic transformations should be learned by rotational flow fields. For coloring, both vector fields

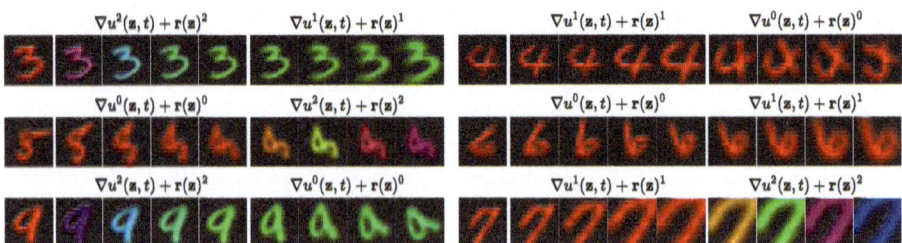

Fig. 7.11 Traversal results of switching latent flows

7.5 Discussions

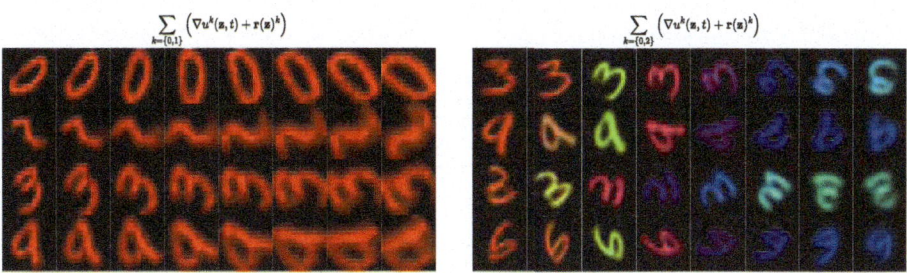

Fig. 7.12 Traversal results of combining latent flows

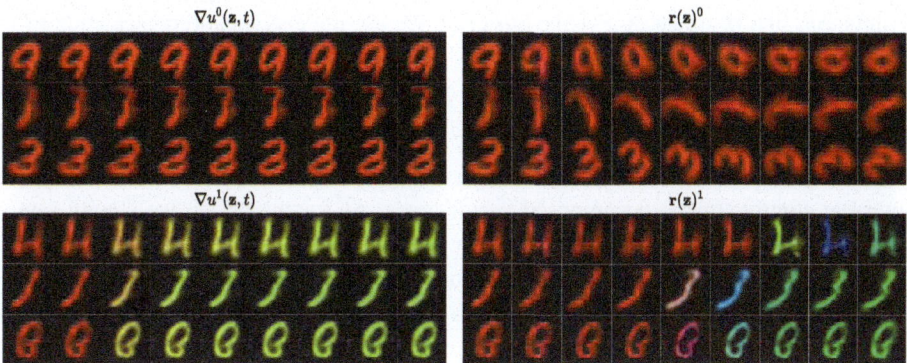

Fig. 7.13 Traversal results using different types of vector fields

are important and contribute to different parts of the transformations. This observation also intuitively makes sense as non-periodic transformations can be learned by both types of vector fields. Interestingly, ∇u^1 mainly manipulates the image in the initial steps while r^1 takes care of the later stage, which implies that the two flow fields can complement each other in different traversal phases.

7.5.3 Learning Separate Controls

With a slight modification to our method, each transformation primitive can be associated with a specific vector field, which could make the Helmholtz decomposition more compelling. To this end, we can introduce separate controls y_{1t}, y_{2t} for the curl-free and divergence-free vector fields:

$$z_t = z_{t-1} + \sum_k \tilde{g}_t^k \left(y_{1t}^k \nabla u^k(z, t) + y_{2t}^k r^k(z) \right) \tag{7.16}$$

Table 7.9 The learned association of different vector fields for each transformation on MNIST

Seed	Scaling	Rotation	Coloring
42	$\nabla u^0(z)$	$r^1(z)$	$r^2(z)$
3857	$\nabla u^0(z) + r^0(vz)$	$r^1(z)$	$r^2(z)$

The above formulation slightly modifies Eq. 7.8 in controlling the sample evolution. The two vector fields therefore share the same speeds while having separate switches. This increases the flexibility of choosing flow fields, thus matching the goal of learning to segregate the symmetries and invariances. For the posterior, we use the analytical representation of the OR gate to compose $\mathbf{y}_t$ as:

$$\mathbf{y}_t = \mathbf{y}_{1t} + \mathbf{y}_{2t} - \mathbf{y}_{1t}\mathbf{y}_{2t} \tag{7.17}$$

This means that if either $\mathbf{y}_{1t}$ or $\mathbf{y}_{2t}$ is active, their 'global' spike variable $\mathbf{y}_t$ will be active. Accordingly, the posterior $q_y(\mathbf{y}_t|\mathbf{x}_t,\mathbf{x}_{t-1})$ is changed to $q_y(\mathbf{y}_{1t},\mathbf{y}_{2t}|\mathbf{x}_t,\mathbf{x}_{t-1})$ to allow for inferring controls of the decomposed vector fields. As for the priors, we simply sample $\mathbf{y}_{1t}, \mathbf{y}_{2t}$ from the candidates $\{10, 01, 11\}$ if $\mathbf{y}_t$ is active.

Table 7.9 displays the vector field correspondences using separate controls with different random seeds. For periodic transformations like rotation, our model learns to associate the flow with a divergence-free vector field. In contrast, the non-periodic transformations are modeled either by a curl-free field alone or by the combination of both flow fields. The results are very coherent with the analysis in Sect. 7.5.2 that the two vector fields play different roles in modeling transformations. Further, the separate control justifies the application of the Helmholtz decomposition in learning latent flows for flexibly modeling input transformations. We do not present the decomposed controls as the main approach because the training can be non-trivial if we further introduce slab variables for speed variations. Nonetheless, we empirically find that this approach works well when there is only the spike component to be modeled.

References

1. Horace B Barlow et al. Possible principles underlying the transformation of sensory messages. *Sensory communication*, 1(01), 1961.
2. Bruno A Olshausen and David J Field. Sparse coding with an overcomplete basis set: A strategy employed by V1? *Vision research*, 37(23):3311–3325, 1997.
3. Pierre Comon. Independent component analysis, a new concept? *Signal processing*, 36(3):287–314, 1994.
4. Laurenz Wiskott and Terrence J Sejnowski. Slow feature analysis: Unsupervised learning of invariances. *Neural computation*, 14(4):715–770, 2002.
5. David Klindt, Lukas Schott, Yash Sharma, Ivan Ustyuzhaninov, Wieland Brendel, Matthias Bethge, and Dylan Paiton. Towards nonlinear disentanglement in natural data with temporal sparse coding. *ICLR*, 2021.

6. H von Helmholtz. Über integrale der hydrodynamischen gleichungen, welche den wirbelbewegungen entsprechen. 1858.
7. H von Helmholtz. Lxiii. on integrals of the hydrodynamical equations, which express vortex-motion. *The London, Edinburgh, and Dublin Philosophical Magazine and Journal of Science*, 1867.
8. Toby J Mitchell and John J Beauchamp. Bayesian variable selection in linear regression. *Journal of the american statistical association*, 1988.
9. Yann LeCun. The mnist database of handwritten digits. 1998.
10. Ralph Abraham, Jerrold E Marsden, and Tudor Ratiu. *Manifolds, tensor analysis, and applications*. Springer Science & Business Media, 2012.
11. Jack Richter-Powell, Yaron Lipman, and Ricky TQ Chen. Neural conservation laws: A divergence-free perspective. *NeurIPS*, 2022.
12. Eric Jang, Shixiang Gu, and Ben Poole. Categorical reparameterization with gumbel-softmax. *ICLR*, 2017.
13. Chris Burgess and Hyunjik Kim. 3d shapes dataset. https://github.com/deepmind/3dshapes-dataset/, 2018.
14. Weili Nie, Tero Karras, Animesh Garg, Shoubhik Debnath, Anjul Patney, Ankit B Patel, and Anima Anandkumar. Semi-supervised stylegan for disentanglement learning. In *ICML*, 2020.
15. Marius Cordts, Mohamed Omran, Sebastian Ramos, Timo Rehfeld, Markus Enzweiler, Rodrigo Benenson, Uwe Franke, Stefan Roth, and Bernt Schiele. The cityscapes dataset for semantic urban scene understanding. In *CVPR*, 2016.
16. Jennifer J Sun, Tomomi Karigo, Dipam Chakraborty, Sharada P Mohanty, Benjamin Wild, Quan Sun, Chen Chen, David J Anderson, Pietro Perona, Yisong Yue, et al. The multi-agent behavior dataset: Mouse dyadic social interactions. *NeurIPS*, 2021.
17. Yue Song, Andy Keller, Nicu Sebe, and Max Welling. Flow factorzied representation learning. In *NeurIPS*, 2023.
18. Yue Song, Andy Keller, Nicu Sebe, and Max Welling. Latent traversals in generative models as potential flows. In *ICML*. PMLR, 2023.
19. T Anderson Keller and Max Welling. Topographic vaes learn equivariant capsules. *NeurIPS*, 2021.
20. Irina Higgins, Loic Matthey, Arka Pal, Christopher Burgess, Xavier Glorot, Matthew Botvinick, Shakir Mohamed, and Alexander Lerchner. beta-vae: Learning basic visual concepts with a constrained variational framework. *ICLR*, 2016.
21. Hyunjik Kim and Andriy Mnih. Disentangling by factorising. In *ICML*, 2018.
22. Diederik P Kingma and Max Welling. Auto-encoding variational bayes. *ICLR*, 2014.
23. Xinqi Zhu, Chang Xu, and Dacheng Tao. Learning disentangled representations with latent variation predictability. In *ECCV*, 2020.
24. Cristina Segalin, Jalani Williams, Tomomi Karigo, May Hui, Moriel Zelikowsky, Jennifer J Sun, Pietro Perona, David J Anderson, and Ann Kennedy. The mouse action recognition system (mars) software pipeline for automated analysis of social behaviors in mice. *Elife*, 2021.

Conclusion 8

8.1 Naturally Inspired Structured Representations

One of the core motivations for the research presented in this book was based on the idea that existing structured representations have clear computational value, but are currently unable to capture the full complexity of the structure present in the natural world. In Part II of this book, we considered alternative generalized forms of structure present in natural neural networks, seeking to determine if natural properties such as topographic organization and spatiotemporal dynamics may fulfill the role of generalized equivariant structure. To do so, we described two methods for integrating structure from natural neural networks into artificial neural networks. Specifically, both methods relied on spatial organization of neurons, and thereby demonstrated a component of 'topographic' organization of both neural selectivity and neural dynamics. Ultimately, we found that indeed, firstly, topographic organization as an inductive bias on its own has the ability to structure neural space according to the co-occurrence statistics of the input data, and thereby can serve as a mechanism for learning structured representations directly from transforming input sequences. Secondly, we found that a bias towards smooth spatiotemporal dynamics, such as traveling waves of neural activity, has a similar potential to organize neural selectivity in a way that allows for causal manipulation of generated outputs, again learned directly from transforming input sequences. We found that these two biases are more intimately related than previously thought, with an observed emergence of topographic organization from a simple bias towards local connectivity and wavelike dynamics. Finally, we found that these sequence transformations can be arbitrary non-group transformations, thereby generalizing the types of structures to which we can build approximately equivariant neural networks, as desired.

In summary, this work has demonstrated the clear potential for natural forms of neural representational structure to benefit learning efficiency and generalization in learning systems of all types, and has highlighted the importance of pursuing this direction further. As evidence of the possibilities in this future direction, a higher level perspective paper has further been written by some of the authors of this book [1], outlining a general framework by which spatiotemporal dynamics in brains and machines may serve as a generalized inductive bias towards symmetries of the data distribution they intend to model, building directly on the insights presented in Part II of this book.

8.2 Learned Homomorphisms and Disentangled Representations

In Part III of the book, we proceeded with an alternative view of how structure with respect to complex natural transformations may be integrated into artificial neural networks. In these chapters, we presented a set of general frameworks inspired by physics, optimal transport, and neuroscience, which can serve to both identify smooth 'disentangled' paths through latent space of pre-trained models corresponding to semantically meaningful transformations in the data space, as well as regularize learning of new models to structure their latent space from random initialization, eventually achieving the desired learned homomorphisms. We present these frameworks as a potential formalized definition of what may be considered 'learned equivariance' in a manner that is far more flexible than existing analytic methods.

This work has led to promising results on the identification of structure in existing large generative models, hinting at a potential promising application to interpretability research for the large image and language models of today; the structures proposed in the book can be translated into diffusion models and Transformers. The work has further demonstrated significant potential in learning disentangled representations of natural transformations from video sequences, with early applications to animal recordings hinting at a potential use for these methods in the natural and biological sciences. In future work, we believe exploring the scientific applications of these methods would be of utmost interest, including applications to modeling neural spiking data, data from astronomical recordings, and data from physical experiments such as particle accelerators. Furthermore, we believe that many of these models could have a more direct connection with neural dynamics, which are also often formally described in terms of potential flows, and thereby may lead to a better understanding of biological neural dynamics broadly.

Reference

1. T. Anderson Keller, Lyle Muller, Terrence J. Sejnowski, and Max Welling. A spacetime perspective on dynamical computation in neural information processing systems, 2024.

The manufacturer's authorised representative in the EU is Springer Nature Customer Service Centre GmbH, Europaplatz 3, 69115 Heidelberg, Germany. If you have any concerns regarding our products, please contact ProductSafety@springernature.com

Printed and bound by CPI Group (UK) Ltd, Croydon, CR0 4YY
26/03/2026
02078941-0016